ACS SYMPOSIUM SERIES **388**

Flavor Chemistry
^
Trends and Developments

Roy Teranishi, EDITOR
Agricultural Research Service
U.S. Department of Agriculture

Ron G. Buttery, EDITOR
Agricultural Research Service
U.S. Department of Agriculture

Fereidoon Shahidi, EDITOR
Memorial University of Newfoundland

Developed from a symposium sponsored
by the Division of Agricultural and Food Chemistry
at the Third Chemical Congress of North America
(195th National Meeting of the American Chemical Society),
Toronto, Ontario, Canada,
June 5–11, 1988

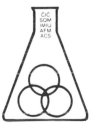

American Chemical Society, Washington, DC 1989

Library of Congress Cataloging-in-Publication Data

Chemical Congress of the North American Continent (3rd: 1988: Toronto, Ont.)

Flavor chemistry: trends and developments
  Roy Teranishi, editor, Ron G. Buttery, editor, Fereidoon Shahidi, editor;
  Developed from a symposium sponsored by the Division
of Agricultural and Food Chemistry at the Third Chemical
Congress of the North American Continent (195th Meeting
of the American Chemical Society), Toronto, Ontario,
Canada, June 5–11, 1988.

  p.   cm.—(ACS Symposium Series, ISSN 0097–6156;
388).
  Includes bibliographies and indexes.

  ISBN 0–8412–1570–7
  1. Flavor—Congresses. 2. Flavoring essences—
Congresses.

  I. Teranishi, Roy, 1922– . II. Buttery, Ron G. III.
Shahidi, Fereidoon, 1951– . IV. American Chemical
Society. Division of Agricultural and Food Chemistry.
V. American Chemical Society. Meeting (195th: 1988:
Toronto, Ont.). VI. Title. VII. Series.

TP372.5.C465   1989
664'.06—dc19   88–35907
                                                   CIP

Foreword

The ACS SYMPOSIUM SERIES was founded in 1974 to provide a medium for publishing symposia quickly in book form. The format of the Series parallels that of the continuing ADVANCES IN CHEMISTRY SERIES except that, in order to save time, the papers are not typeset but are reproduced as they are submitted by the authors in camera-ready form. Papers are reviewed under the supervision of the Editors with the assistance of the Series Advisory Board and are selected to maintain the integrity of the symposia; however, verbatim reproductions of previously published papers are not accepted. Both reviews and reports of research are acceptable, because symposia may embrace both types of presentation.

Contents

Preface

Food is one of the most intimate and important components of our chemical environment. Whether we accept or reject food depends mainly on its flavor. Research into the chemistry of desirable and undesirable foods has become very popular; the advent of modern instrumentation has introduced rapid changes in the field.

The sequence of emphasis in flavor chemistry research has been the following:

1. Experimental methods, or how to obtain the information we need. This aspect must continue as problems being studied become more complex.

2. Correlation of chemical structure to sensory properties. Research into this topic must also be continued. Some simple problems have been solved, and now more complex flavors are being elucidated.

3. Formation origin and mechanisms of flavors. This has always been a topic of interest, but with more definite information available about the chemical composition of characteristic flavors, more definite paths of biological and chemical origin can be postulated and verified.

Understanding the mechanism by which flavor compounds are formed can lead to better methods of food processing for better formation and retention of flavor. Fundamental flavor chemistry information is also essential in genetic engineering of plants and animals to improve flavor in the starting materials of food products.

To illustrate some trends and developments in flavor chemistry research, chapters have been included on the importance of enantiomers and the biogenesis of some chiral compounds, production and mechanisms of natural and chemically formed flavor compounds, and a few recent examples of chemical investigations of characteristic flavors.

Ron G. Buttery
Western Regional Research Center
Agricultural Research Service
U.S. Department of Agriculture
Albany, CA 94710

FEREIDOON SHAHIDI
Department of Biochemistry
Memorial University of Newfoundland
St. Johns's, Newfoundland A1B 3X9, Canada

ROY TERANISHI
Western Regional Research Center
Agricultural Research Service
U.S. Department of Agriculture
Albany, CA 94710

November 1988

Chapter 1

New Trends and Developments in Flavor Chemistry

An Overview

Roy Teranishi

Western Regional Research Center, Agricultural Research Service, U.S. Department of Agriculture, Albany, CA 94710

This chapter gives an overview of new trends and developments in flavor chemistry. One important development was made possible by advances in analytical methodology, that is, the identification of numerous compounds with known flavor characteristics. As more and more compounds are correlated with characteristic flavors, there is a trend to study flavor precursors and to explain how flavor is developed and released. Many of the newest developments in flavor chemistry are in the area of flavor production from plant and animal sources; this trend has come about because of the public's fear of the words "chemical" and "synthetic". In this chapter, words such as these are discussed in terms of the public's perception of them versus a chemist's viewpoint. Another new trend is to understand the chemical reactions involved in the processing and storage of foods in order to bring foods to consumers at optimum acceptability.

Prior to the 1950's only about 500 flavor compounds were known (1). Since then, with the advent of modern instrumention, thousands of compounds have been characterized in hundreds of different foods (2). There have been many books published on flavor research workshops and symposia, some of which are held on a periodic basis and some on special occasions and topics, covering various aspects of flavor (3-20). Also, there are many excellent reviews which every serious flavor chemist should consult (21-43).

Advances in analytical methodology introduced in the 1960's were applied from the early 1970's (3, 5, 12, 13, 14). Previous to gas chromatography, fractional distillations and column chromatography of colored derivatives were the primary means of separations. Size of sample required for distillation is, of course, enormous compared to what is required for gas chromatography. The resolution of separation by gas chromatography is far superior to that attained by

fractional distillation. Also, the advent of infrared, nuclear magnetic resonance, and mass spectrometry has made it possible to make
structural determinations with micro amounts. Thus, with the use of
modern analytical methods, the number of compounds with known flavor
characteristics increased in the 1970's and 1980's. These advances
have set the stage for the present trends and developments in flavor
chemistry.

Many of the newest developments in flavor chemistry are in the area
of flavor production from plant and animal sources; hence, signi-
fying the popularity of the term "biotechnology". This trend has
come about because of the fear of the public of the words "chemical"
and "synthetic".

The word "natural" is used in opposition to the word "synthetic"
with the connotation that "natural" products are safer than "synthetics", but there are many toxins made by plants and animals which
are very detrimental to man (44-47). Moreover, whether chemicals
are made in flasks by man or made by plants and animals, no
compounds are made on earth other than those permitted by the laws
of nature. Therefore, all molecules on this earth are "natural".
This definition is from a chemist's viewpoint, and is in agreement
with a dictionary definition of "characteristic of or explainable by
the operations of the physical world".

However, The Food and Drug Administration definitions (48) are:
"The term 'artificial flavor' or 'artificial flavoring' means any
substance, the function of which is to impart flavor, which is not
derived from a spice, fruit or fruit juice, vegetable or vegetable
juice, ... , or fermentation products thereof. ... The term
'natural flavor' or 'natural flavoring' means the essential oil,
oleoresin, essence or extractive, protein hydrolysate, distillate or
any product of roasting, heating or enzymolysis, which contains the
flavoring constituents derived from a spice, fruit or fruit juice,
vegetable or vegetable juice, edible yeast, herb, bark, bud, root,
leaf or similar plant material, meat, seafood, poultry, eggs, dairy
products, or fermentation products thereof." It is this set of
words, "or fermentation products thereof", which has set off a
flurry of activity in biotechnology in order to use the words
"natural flavoring" on the label of food products.

The public has an unbased fear of "chemicals" and "synthetics" and an
unbased confidence in "natural" compounds and products. The public
should be educated that there are no differences in the molecules
used in flavorings which are made by man in flasks or by plants and
animals. However, because of this fear, there is a trend in the use
of "naturals", materials from plants and animals (including microorganisms) obtained by biotechnology, over "synthetics", materials
from chemical laboratories.

Plants and animals have been selected by classical genetic methods
for optimum yield, color, texture, disease resistance, etc. It is
time for plants and animals to be selected for optimum flavor. In

this long range plan, flavor chemistry will help in the selection of plants and animals for acceptability whether it is by the classical or modern genetic engineering methods.

In the short range plan, raw materials can be harvested, processed, stored and shipped to bring more flavorful food products to consumers. Very few food products are used directly as grown on the farm. Grains must be milled, made into flour, and then baked to make breads and other cereal products. Fruits and vegetables must be picked at a time to give optimum flavor and texture. Most meats, red or white flesh, are almost flavorless until heated. Chemical reactions involved in the above situations must be understood in order to bring foods to consumers at optimum acceptability.

The development of modern analytical methods has permitted the examination of volatiles from fresh fruit to determine when to pick the fruit. It is the usual concept that fruits are at their best when picked "tree-ripe". However, in extreme cases, as with bananas and pears, these fruits must be picked when hard and green and be permitted to soften and ripen off the tree. If these fruits are permitted to ripen on the tree, they become mealy and unacceptable. Some fruit, like strawberries and peaches, are of best quality when left to ripen on the plants. Apples have been shown to have the best aroma if picked almost ripe and develop the most aroma about a week or two after picking.

Man has used biotechnology for converting raw materials to food products for many centuries. Production of beer is thought to date back to about 6000 BC in ancient Babylonia. The predecessors of soy sauce and miso seem to have originated in China some 2500 years ago. Man has been using products altered by microorganisms, and has learned by trial and error which of the products are safe to eat and which are not. Modern scientific methods are now being applied to explain what chemical and physical alterations are accomplished by microorganisms. Also, further advancements in biotechnology will probably be made to make such systems even more efficient, perhaps even bypass living organisms by using only enzymes. As more and more compounds are correlated with characteristic flavors, there is a trend to study flavor precursors and to explain how flavor is developed and released, especially since now it can be determined exactly which enantiomer is making the contribution to a characteristic aroma.

The era of publishing a large number of compounds identified as to chemical structures is slowly changing to an era in which constituents are identified as to which are the important contributors to the characteristic odors. More and more sensory analyses are stating odor threshold values as well as odor quality.

In the evaluation of contribution to taste, amino acids and peptides are being studied as to sweet, salty, bitter, sour and umami [brothy mouth-feel, see (19)] sensations. In the production of gravies and soups, proteins are hydrolyzed to smaller molecules which evoke more taste sensations than do the large protein molecules. Systematic studies of amino acids and peptides are providing interesting data which will be useful in optimizing conditions to yield the greatest amount of acceptable tastes and minimizing undesirable tastes.

As analytical methodology is improved, the known number of compounds contributing to flavor will be increased, and flavor chemistry will become more applied. Industrial organizations will be able to utilize the information gained in fundamental research to improve the quality of their products. At the same time, there will be more of a data base on which to build a better understanding of the mechanisms of perception of taste and olfaction.

As the demand for natural flavors increases, and as constituents contributing to such flavors are identified, flavor chemistry will be applied in the biotechnological production of such flavors. Also, processing methods will be followed to retain most of the fresh flavors of raw materials. Cases in which flavor is developed during processing, modern analytical methods will be applied to adjust processing conditions to produce the optimum desirable flavors. Thus, flavor chemistry has reached a stage where it is now being applied to improve the flavor of foods, fresh and processed, reaching many consumers.

Literature Cited

1. Weurman, C. Lists of Volatile Compounds in Foods, 1st Edition; Division of Nutrition and Food Research TNO: Zeist, The Netherlands, 1963.
2. Volatile Compounds in Food; S. van Straten and H. Maarse, Ed.; Division of Nutrition and Food Research TNO: Zeist, The Netherlands, 1983. Supplement 1, 1984; 2, 1985; 3, 1986; 4, 1987.
3. Symposium on Foods: The Chemistry and Physiology of Flavors; H. W. Schultz, E. A. Day, and L. M. Libbey, Ed.; AVI: Westport, Connecticut, 1967; 552 pp.
4. Gustation and Olfaction; G. Ohloff and A. F. Thomas, Ed.; Academic Press: London, 1971; 274 pp.
5. Teranishi, R., I. Hornstein, P. Issenberg, and E. L. Wick. Flavor Research: Principles and Techniques; Marcel Dekker, Inc.: New York, 1971; 315 pp.
6. Aroma Research; H. Maarse and P. J. Groenen, Ed.; Pudoc: Wageningen, 1975; 245 pp.
7. Geruch- and Geschmackstoffe; F. Drawert, Ed.; H. Carl: Nurnberg, 1975; 299 pp.
8. Phenolic, Sulfur, and Nitrogen Compounds in Food Flavors; G. Charalambous and I. Katz, Ed.;ACS Symposium Series 26. American Chemical Society: Washington, DC, 1976; 215 pp.
9. Progress in Flavour Research; D. G. Land and H. E. Nursten, Ed.; Applied Science Publishers, Ltd.: London, 1979; 371 pp.
10. Food Taste Chemistry; J. C. Boudreau, Ed.; ACS Symposium Series 115, ACS: Washington, DC, 1979; 262 pp.
11. Flavour '81; P. Schreier, Ed.; de Gruyter: Berlin, 1981; 780 pp.

12. Flavor Research: Recent Advances; R. Teranishi, R. A. Flath, and
    H. Sugisawa, Ed.; Marcel Dekker, Inc.: New York, 1981; 381 pp.
13. Analysis of Volatiles; P. Schreier, Ed.; de Gruyter: Berlin,
    1984; 469 pp.
14. Analysis of Foods and Beverages: Modern Techniques; G.
    Charalambous,Ed.; Academic Press: New York, 1984; 652 pp.
15. Progress in Flavour Research 1984; J. Adda, Ed.; Elsevier:
    Amsterdam, 1985.
16. Topics in Flavour Research; R. G. Berger, S. Nitz, and P.
    Schreier, Ed.; H. Eichorn: D-8051 Marzling-Hangenham, 1985; 476
    pp.
17. Chemical Changes in Food during Processing; T. Richardson and J.
    W. Finley, Ed.; AVI: Westport, Connecticut, 1985; 514 pp.
18. Chemistry of Heterocyclic Compounds in Flavours and Aromas; G.
    Vernin, Ed.; Ellis Horwood, Ltd.: Chichester, 1982; 375 pp.
19. Umami: A Basic Taste; Y. Kawamura and M. R. Kare, Ed.; Marcel
    Dekker, Inc.: New York, 1987; 649 pp.
20. Flavour Science and Technology; M. Martens, G. A. Dalen and H.
    Russwurm, Jr., Ed.; Wiley: London, 1987; 566 pp.
21. Ohloff, G. Importance of minor components in flavors and
    fragrances. Perfumer and Flavorist 1978, 3, 11.
22. Ohloff, G., Recent developments in the field of naturally-
    occurring aroma components. In Progress in the Chemistry of
    Organic Natural Products, 1978, Vol. 35, p. 431, (founded by L.
    Zechmeister) W. Herz, H. Griseback, G. W. Kirby, Ed.;
    Springer-Verlag: Wien - New York.
23. Ohloff, G., and I. Flament. Some recent aspects of the
    chemistry of naturally occuring pyrazines. In The Quality of
    Foods and Beverages. Chemistry and Technology. Vol. 1, G.
    Charalambous and G. Inglett, Ed.; Academic Press: New York,
    1981; p. 35.
24. Ohloff, G., I. Flament,, and W. Pickenhagen. Flavor chemistry,
    Food Reviews International, 1985, 1(1): 99.
25. Maga, J. A., and C. E. Sizer. Pyrazines in foods, Handbook of
    Flavor Ingredients, 2nd Edition, vol. 1; CRC Press: Cleveland,
    1975, p. 47.
26. Maga. J. A. Thiazoles in foods, ibid., p. 228.
27. Maga, J. A. Bread flavor, ibid., p. 669.
28. Maga, J. A. The role of sulfur compounds in food flavor. Part
    I. Thiazoles, CRC Crit. Rev. Food Sci. Nutr., 1975, 6(2): 153.
29. Maga, J. A., Part II. Thiophenes, ibid., 241.
30. Maga, J. A., Part III. Thiols, ibid., 1976,7(2): 147.
31. Maga, J. A., Lactones in food, ibid., 8(1): 1.
32. Maga, J. A., Phenolics in food, ibid., 1978,10(4): 323.
33. Maga, J. A., Amines in food, ibid., 10(4): 373.
34. Maga, J. A., Furans in food, ibid., 1979,11(4): 355 .
35. Maga, J. A., The chemistry of oxazoles and oxazolines in food,
    ibid., 1981, 14(3): 285.
36. Maga, J. A., Pyrazines in foods: an update, ibid., 1982, 16: 1.
37. Maga, J. A., Flavor potentiators, ibid., 1984, 18: 231 .
38. Maga, J. A., The flavor chemistry of wood smoke, Food Reviews
    International, 1987, 3(1 & 2): 139.
39. Belitz, H.-D., and H. Wieser. Bitter compounds: occurrence and
    structure-activity, ibid., 1985, 1(2): 271.

40. Petro-Turza, M.   Flavor of tomato and tomato products, ibid.,
    1986–1987, 2(3): 309.
41. Carson, J. F.   Chemistry and biological properties of onions and
    garlic, ibid., 1987, 3(1 & 2): 71.
42. Fukushima, D.   Fermented vegetable protein and related foods of
    Japan and China, ibid., 1985, 1(1): 149.
43. Bioflavour '87; edited by P. Schreier, in press.
44. Toxic Constituents of Plant Foodstuffs; I. E. Liener, Ed.;
    Academic Press: New York, 1969; 500 pp.
45. Hirono, I.   Natural carcinogenic products of plant origin, CRC
    Crit. Rev. Toxicol.   1981, 8(3), 235–277.
46. Nutritional and Toxicological Aspects of Food Safety;   M.
    Friedman, Ed.; Plenum Press: New York, 1984; 584 pp.
47. Plant Toxicology; M. P. Hegarty, L. F. James, R. F. Keeler, Ed.;
    Dominion Press: Melbourne, 1985; 623 pp.
48. Code of Federal Regulations, Food and Drugs, Vol. 21, Part
    101.22, Office of the Federal Register, National Archives and
    Records Administration, U. S. Government Printing Office:
    Washington, D. C., April, 1988.

RECEIVED August 30, 1988

FORMATION OF FLAVOR COMPOUNDS

Chapter 2

Biosynthesis of Chiral Flavor and Aroma Compounds in Plants and Microorganisms

K.-H. Engel, J. Heidlas, W. Albrecht, and R. Tressl

Technische Universität Berlin, Fachbereich Lebensmitteltechnologie und Biotechnologie, Fachgebiet Chemish-technische Analyse, Seestr.13, D–1000 Berlin 65, Federal Republic of Germany

Capillary gas chromatographic determination of optical purities, investigation of the conversion of potential precursors,
and characterization of enzymes catalyzing these reactions
were applied to study the biogenesis of chiral volatiles in
plants and microorganisms. Major pineapple constituents are
present as mixtures of enantiomers. Reductions, chain elongation, and hydration were shown to be involved in the biosynthesis of hydroxy acid esters and lactones. Reduction of
methyl ketones and subsequent enantioselective metabolization
by Penicillium citrinum were studied as model reactions to
rationalize ratios of enantiomers of secondary alcohols in
natural systems. The formation of optically pure enantiomers
of aliphatic secondary alcohols and hydroxy acid esters using
oxidoreductases from baker's yeast was demonstrated.

The world-wide trend to "natural" flavor and aroma has significantly
increased interest in biogenetical pathways leading to volatiles in
natural systems. For chiral compounds the exploration of potential
biosynthetic routes is even more important, because chemical syntheses are often difficult and expensive; in many cases however sensory qualities of enantiomers are different (1-3). In our current
studies of chiral volatiles in plant and microbial systems we use
different analytical approaches. (a) Capillary gas chromatographic
separations of diastereoisomeric derivatives are used to determine
the configurations of chiral constituents at trace levels. (b) Chemically synthesized (labeled) precursors are added to fruit tissues and
microorganisms. Their transformation into chiral constituents is
investigated by means of capillary gas chromatography/mass spectrometry; the stereochemical course of these metabolizations is followed. (c) Enzymes catalyzing the stereospecific conversion of precursors to chiral compounds are isolated and characterized; commercially available enzymes are investigated as model systems to elucidate the stereochemical course of biogenetical pathways. The combination of these methods revealed some new aspects of the biosynthesis
of chiral compounds in natural systems.

0097–6156/89/0388–0008$06.00/0

Naturally Occurring Configurations of Pineapple Volatiles

A distinct feature of the spectrum of volatiles isolated from pineapple (Ananas comosus (L.) Merr.) is the presence of numerous chiral
components: 3- and 5-hydroxy esters, 3-, 4- and 5-acetoxy esters,
and γ- and δ-lactones are prominent pineapple flavor and aroma
constituents (4-7). Capillary gas chromatographic separation of
diastereoisomeric derivatives of (S)-(+)-α-methoxy-α-trifluoro-
methylphenylacetic acid chloride (MTPA) and (R)-(+)-phenylethyliso-
cyanate (PEIC) revealed that these chiral pineapple components are
not contained in optically pure form, but as mixtures of enantiomers.
The enantiomeric ratios listed in Table I confirmed findings of
previous investigations of pineapples (8). Ratios of enantiomers
rather than optically pure constituents have also been observed in
other fruits, such as passion fruit and mango (9).

Table I. Concentrations and enantiomeric compositions of chiral
constituents in firm-mature and soft-ripe pineapples

compound	concentration (ppb)		increase (%)[a]	enantiomeric composition[b]			
	firm-mature	soft-ripe		firm-mature %(S) %(R)		soft-ripe %(S) %(R)	
methyl 3-hydroxy-hexanoate	310	390	26	84	16	86	14
methyl 3-acetoxy-butanoate	120	380	217	-c		83	17
methyl 3-acetoxy-hexanoate	1630	3680	126	91	9	93	7
methyl 4-acetoxy-hexanoate	480	540	13	73	27	75	25
methyl 5-acetoxy-hexanoate	1200	1610	34	64	36	62	38
methyl 4-acetoxy-octanoate	-	<30	-			-c	
methyl 5-acetoxy-octanoate	1000	1050	5	-b		51	49
γ-hexalactone	630	1360	116	76	24	80	20
δ-hexalactone	310	320	3	16	84	37	63
δ-octalactone	750	580	-	49	51	53	47

a) ripening over 5 days at room temperature;
b) determined by GC-investigation of MTPA- and PEIC-derivatives,
   respectively;
c) enantiomeric composition not determined.

The biogenesis of volatiles in plant systems is a very dynamic
process. Changes in the activities of enzymes and pathways at differ-
rent stages of maturity and ripeness might influence not only the
concentrations but also the optical purities of chiral constituents.
Therefore we subjected "firm-mature" pineapples to post-harvest
ripening at room temperature over a period of 5 days. The concentrations of all chiral components (except δ-octalactone) increased
significantly. The enantiomeric compositions, however, remained nearly unchanged during this ripening process (Table I). This constancy
of enantiomeric ratios at different physiological stages of the
fruits is a very important result. The independence of optical purities on maturity and ripeness is one of the premises to use the
investigation of chiral constituents for detection of adulterations
in fruit products (10).

Addition of Oxocompounds to Pineapple Tissue

Mixtures of enantiomers rather than optically pure compounds can be
rationalized by different assumptions: (a) one enzyme with low enantioselectivity catalyzes the biogenetical process, (b) at least two
enzymes with different enantioselectivities compete in the reaction,
and (c) different pathways eventually leading to the same final products, however with opposite configurations, are involved. By addition of chemically synthesized precursors we aimed to trace some of
the biogenetical routes involved in the biogenesis of chiral pineapple volatiles.
The distribution of metabolites obtained after incubation of pineapple slices with keto acids and keto esters, potential precursors of
the corresponding hydroxy compounds, is summarized in Table II. The
metabolization steps comprise esterification, reduction to hydroxy
compounds, formation of acetoxy esters, and cyclization to the corresponding lactones. Metabolization rate and distribution of formed
products strongly depend on the structures of the precursors. The detection of these metabolites proves the enzymatic capability of pineapple tissue to catalyze these conversions, an aspect which might be
interesting for future use of pineapple tissue cultures in the production of chiral compounds.
Capillary gas chromatographic investigation of diastereoisomeric
derivatives revealed that in analogy to results obtained without
precursors the chiral metabolites are present as mixtures of enantiomers. However for only a few of these compounds the ratios of enantiomers are identical with those determined in pineapple without
precursors. The enantiomeric compositions of ethyl 3-hydroxyhexanoate
and ethyl 3-acetoxyhexanoate are almost opposite to those determined
for the naturally occurring methyl esters. δ-Octalactone obtained
after addition of 5-oxooctanoic acid to pineapple tissue is almost
optically pure (92% S); on the other hand δ-octalactone is naturally
present in pineapple tissue as nearly racemic mixture (Table I,8).
One might conclude from these differences in the optical purities that the reduction of ketoprecursors can not be the major pathway active in the biosynthesis of hydroxy esters, acetoxy esters,
and lactones in pineapple. However, when interpreting these results
it must be considered that in the case of a competition of two en-

Table II. Concentrations and enantiomeric compositions of metabolites formed after incubation of pineapple slices

precursor products	concentration (ppb) before after incubation[a]		degree of metabo- lization (%)	enantiomeric composition[b] %(S) %(R)	
ethyl 3-oxohexanoate			85		
ethyl 3-hydroxyhexanoate	–	34540		18	82
ethyl 3-acetoxyhexanoate	–	5850		31	69
ethyl 4-oxohexanaote			12		
ethyl 4-hydroxyhexanoate	–	3200		11	89
ethyl 4-acetoxyhexanoate	–	180		–[c]	
γ-hexalactone	300	2380		23	77[d]
ethyl 5-oxohexanoate			23		
ethyl 5-hydroxyhexanoate	–	7250		67	33
ethyl 5-acetoxyhexanoate	–	1950		76	24
δ-hexalactone	40	3050		60	40[d]
methyl 5-oxohexanoate			9		
methyl 5-hydroxyhexanoate	<20	4670		50	50
methyl 5-acetoxyhexanoate	430	1750		85	15
δ-hexalactone	50	1090		44	56
5-oxohexanoic acid			2.5		
methyl 5-oxohexanoate	–	2440		–	
methyl 5-hydroxyhexanoate	<20	1960		27	73
methyl 5-acetoxyhexanoate	319	457		–[c]	
δ-hexalactone	56	1643		40	60[d]
5-oxooctanoic acid			3		
methyl 5-oxooctanoate	–	500		–	
methyl 5-hydroxyoctanoate	<10	540		68	37
methyl 5-acetoxyoctanoate	150	160		–[c]	
δ-octalactone	520	6920		8	92[d]
5-oxodecanoic acid			4		
methyl 5-oxodecanoate	–	140		–	
methyl 5-hydroxydecanoate	–	–		–	
methyl 5-acetoxydecanoate	–	–		–	
δ-decalactone	–	13800		20	80[d]

a) incubation time: 20 hours; concentration of precursors: oxoesters: 0.4 mmol/kg; oxoacids: 2.0 mmol/kg;
b) determined by GC-investigation of MTPA- and PEIC-derivatives, respectively;
c) enantiomeric composition not determined;
d) enantiomeric composition determined by GC-investigation of PEIC-derivatives of the corresponding diols.

zymes, showing different enantioselectivities and different K_m-values
for a common substrate, the concentration of this substrate decisively determines the reaction rates of the enzymes and thus the enantiomeric composition of the product. Therefore the "unnaturally" high
concentration of a precursor, as a result of the addition of a relatively high amount of the chemically synthesized compound to the
plant tissue, may be the reason that optical purities of products are
different from those without precursors.

The presence of at least two enzymes with different enantioselectivities depending on the structures of the substrates, comparable to
the oxoacid ester reductases in baker's yeast (11), is indicated by
the different optical purities obtained after incubation of 5-oxo-
hexanoic acid, methyl 5-oxohexanoate, and ethyl 5-oxohexanoate (Table
II). Variations of the concentrations of added precursors also resulted in different optical purities of metabolites (12).

Addition of Labeled Precursors to Pineapple Tissue

If unlabeled precursors are added the investigation of pathways is
mainly based on quantitative data and requires significant increases
of metabolites. The use of deuterium labeled compounds offers the
possibility to follow conversions of precursors with lower metabolization rates, because the formed products can be detected based on
their content of deuterium by using the more sensitive method of capillary gas chromatography-mass spectrometry. This technique revealed
that two independent pathways starting from different precursors
contribute to the formation of δ-octalactone in pineapple.

Figure 1 shows part of a reconstructed ion chromatogram of a
pineapple aroma extract isolated after incubation of pineapple slices
with 3-hydroxyhexanoic acid-3-d_1. GC-MS detection of deuterated compounds showed that the following pathways are active: (a) esterification leading to methyl and ethyl esters, (b) dehydration to (E)-2-
and (E)-3-hexenoates, and (c) chain elongation to methyl 5-hydroxy
octanoate followed by acetylation (methyl 5-acetoxyoctanoate) and
cyclization (δ-octalactone).

On the other hand deuterated δ-octalactone was also detected
after addition of (Z)-4-octenoic acid-2,2-d_2 to pineapple tissue (Figure 2). Hydration of the double bond followed by cyclization of the
intermediates 4-hydroxyoctanoic acid-2,2-d_2 and 5-hydroxyoctanoic
acid-2,2-d_2 leads to γ-octalactone-2,2-d_2 and δ-octalactone-2,2-d_2,
respectively.

The formation of δ-octalactone by three independent pathways
(a) reduction of 5-oxoprecursors, (b) chain elongation of 3-hydroxy-
hexanoate, and (c) hydration of (Z)-4-octenoic acid demonstrates the
complexity of the biosynthesis of chiral compounds in natural systems.

Oxidoreductases Catalyzing the Enantioselective Reduction of Oxoacid
Esters in Baker's Yeast

Early investigations of reactions catalyzed by Saccharomyces cerevisiae revealed that the stereochemical course of the reduction of
3-oxoacids is influenced by the chain lengths of the substrates:

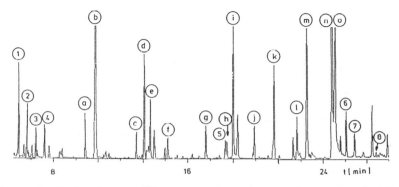

Figure 1. Part of a capillary gas chromatogram (reconstructed ion detection) of an aroma extract of pineapple tissue, isolated after addition of 3-hydroxyhexanoic acid-3-d₁ (CP Wax 52 CB column, 50 m x 0.32 mm i.d., df = 0.25 μm; temperature program 70-220 °C; 4 °C/ min).

(1) ethyl hexanoate-3-d₁
(2) methyl (Z)-3-hexenoate-3-d₁
(3) methyl (Z)-2-hexenoate-3-d₁
(4) ethyl (E)-3-hexenoate-3-d₁
(5) ethyl 3-hydroxyhexanoate-3-d₁
(6) methyl 5-acetoxyoctanoate-5-d₁
(7) δ-octalactone-5-d₁
(8) methyl 5-hydroxyoctanoate-5-d₁

(a) methyl octanoate
(b) octanol-2 (internal standard)
(c) dimethyl malonate

(d) methyl 3-methylthiopropionate
(e) methyl 3-acetoxybutanoate
(f) O-methyl furaneol
(g) methyl 3-hydroxyhexanoate
(h) γ-hexalactone
(i) methyl 3-acetoxyhexanoate
(j) 3-methylthiopropanol-1
(k) methyl 4-acetoxyhexanoate
(l) δ-hexalactone
(m) methyl 5-acetoxyoctanoate
(n) ß-phenylethanol
(o) γ-octalactone

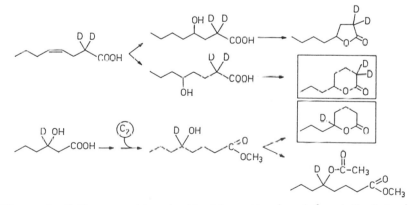

Figure 2. Pathways active in the biosynthesis of δ-octalactone in pineapple.

3-oxobutanoic acid is reduced to (S)-3-hydroxybutanoic acid (13),
whereas the reduction of 3-oxohexanoic acid leads to the (R)-configu-
rated hydroxyacid (14). Sih et al. (15) demonstrated that the optical
purities of 3-hydroxyacid esters obtained by yeast catalyzed reduction can be controlled by variation of the structures of the educts
(e.g. by different chain lengths of acid and alcohol moieties, respectively, of the 3-oxoacid esters). They postulated the existence
of two enzymes with different enantioselectivities and different substrate specificities, the so-called (R)- and (S)-enzymes, which are
competing for the substrate.

By means of streptomycin sulfate treatment, Sephadex G-25 filtration, DEAE-Sepharose CL-6B chromatography, Sephadex G-150 filtration,
and hydroxyapatite chromatography we succeeded in isolating and puri-
fying two NADPH-dependent oxidoreductases from enzyme extracts of
Saccharomyces cerevisiae, which catalyze the enantioselective reduction of 3-oxoacid esters to (S)- and (R)-3-hydroxyacid esters (11).
The elution pattern of enzymes reducing ethyl 3-oxohexanoate obtained after DEAE-Sepharose-CL-6B chromatography is shown in Figure 3.

The (S)-enzyme (relative molecular mass 48000 -50000) reduced
3-oxoacid esters, 4-oxo and 5-oxoacids and esters enantioselectively
to (S)-hydroxy compounds. The K_m-values for ethyl 3-oxobutanoate,
ethyl 3-oxohexanoate , 4-oxopentanoic acid and 5-oxohexanoic acid
were determined as 0.9 mM, 5.3 mM, 17.1 mM and 13.1 mM, respectively.

The (R)-enzyme (molecular mass 800000) was shown to be identical
with a subunit of the fatty acid synthase complex. It reduced 3-oxo-
acid esters specifically to (R)-hydroxyacid esters. K_m-values for
ethyl 3-oxobutanoate and ethyl 3-oxohexanoate were determined as
17.0 mM and 2.0 mM, respectively. Intact fatty acid synthase showed
no activity in catalyzing the reduction of 3-oxoacid esters.

Opposite enantiomers of 3-hydroxyacid esters of different chain
lengths in fruits, such as yellow passion fruit (16) and the influence of the structures of oxoprecursors on the optical purities of 3-
hydroxyacid derivatives in incubation experiments with pineapple indicate a competition of oxidoreductases in plant systems comparable
to baker's yeast.

Secondary Alcohols in Plants

Capillary gas chromatographic investigation of diastereoisomeric derivatives revealed that in some fruits, such as passion fruits (9)
and blackberries (17), secondary alcohols and their esters are contained in almost optically pure form. On the other hand corn (Zea
mays) contains aliphatic secondary alcohols as mixtures of enantiomers; the ratios depend upon the chain lengths of the alcohols.
Heptan-2-ol is present mainly as (R)-enantiomer; with increasing
chain length the proportion of (S)-enantiomer increases. A similar
distribution has been determined in coconut (Figure 4).

To elucidate these results we decided to study the stereochemical
course of the formation of secondary alcohols by microorganisms and
purified enzymes as model systems.

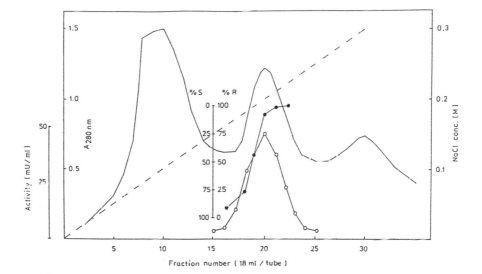

Figure 3. Elution pattern of enzymes reducing ethyl 3-oxohexanoate on DEAE-Sepharose CL-6B chromatography with increasing NaCl concentration. (O) enzyme activity; assay: 1.5 ml 200 mM Tris-HCl buffer (pH 7.2), 0.2 ml 1 mM NADPH solution, 0.1 ml 50 mM ethyl 3-oxohexanoate (emulsified in an aqueuos solution of 5 % propylenglycol) and 0.05 ml enzyme solution; measuring the absorbance at 340 nm at 25 °C. (●) enantiomeric composition (GC separation of MTPA-derivatives) of the formed ethyl 3-hydroxyhexanaote; assay: active fractions (tubes 16 to 23), ethyl 3-oxohexanoate (2 mg) and NADPH (1.5 mg) in a total volume of 3 ml 0.1 Tris-HCl buffer, pH 7.2, for 6 hours. (—) absorbance 280 nm. (- - -) concentration of NaCl. (Reproduced with permission from Ref.11. Copyright 1988 FEBS.)

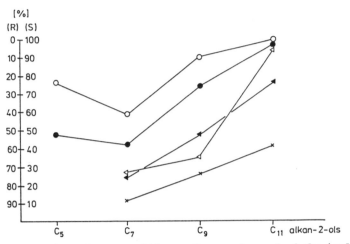

Figure 4. Enantiomeric compositions of secondary alcohols isolated from corn (**X** = Golden Jubilee, **Δ** = Golden Bantam, **▲** = Bonanza) and coconut (**●**) and obtained after reduction of methyl ketones in a shake culture of Penicillium citrinum after 68 h (**O**).

Stereochemical Course of the Formation of Secondary Alcohols by Penicillium citrinum

The formation of methyl ketones can be rationalized by a modified β-oxidation of fatty acid precursors (18). This process, caused by some moulds, mainly deteriorates short chain fatty acids and leads to the so-called "ketonic-rancidity". Kinderlerer et al. (19-21) demonstrated the formation of methyl ketones and secondary alcohols by some xerophilic fungi isolated from coconuts. After deacylation and decarboxylation of the β-ketoacyl-CoA-intermediates the last step of the biogenetical sequence is the reduction of the corresponding methyl ketones.

To study the stereochemical course of this reduction we added a homologous series of aliphatic methyl ketones (C_7-C_{11}) as precursors to a shake culture of Penicillium citrinum. Results obtained for heptan-2-one and nonan-2-one are summarized in Table III. The degree of reduction depends on the chain lengths; maximum conversion was observed for nonan-2-one. Penicillium citrinum preferentially cata-lyzes the formation of (S)-alkan-2-ols. However the optical purities of the formed alcohols were also influenced by the chain lengths. The highest optical purity was obtained for decan-2-ol (96 % enantiomeric excess, e.e. (S)). With decreasing chain length the proportion of (R)-alcohol increased; heptan-2-ol was isolated with a maximum purity of only 64% (S):36% (R). These results indicate the presence of at least two chain length specific oxidoreductases with different enan-tioselectivities comparable to the oxoester reductases in baker's yeast.

Table III: Quantitative distribution and enantiomeric compositions of secondary alcohols obtained by reduction of methyl ke-tones with Penicillium citrinum

reaction-time (h)	heptan-2-ol			nonan-2-ol		
	concentration (ppm)	enantiomeric composition[a] %(S)	%(R)	concentration (ppm)	enantiomeric composition[a] %(S)	%(R)
4	1.1	58	42	16.2	96	4
7	2.5	59	41	22.5	97	3
11	4.9	60	40	24.5	97	3
21	6.7	63	37	23.0	97	3
32	6.9	64	36	20.2	96	4
46	6.7	63	37	14.2	94	6
69	5.5	60	40	4.5	88	12
96	2.2	42	58	0.1	56	44

a) determined by GC investigation of MTPA- and PEIC-derivatives

There is however a second phenomenon decisively influencing the optical purities of the alcohols formed in the course of Penicillium citrinum catalyzed reduction.The formed alcohols are metabolized again; this metabolization proceeds enantioselectively. The preferentially formed (S)-enantiomer is preferentially metabolized. As shown in Table III the optical purity (% enantiomeric excess, e.e.) of nonan-2-ol decreases from 92% e.e.(S) to 12% e.e. (S). Heptan-2-ol is finally present mainly as (R)-enantiomer. The metabolization steps are currently under investigation ; one of the pathways is a hydroxylation leading to hydroxy ketones and diols. Figure 5 presents structures of hydroxylated metabolites obtained from nonan-2-one.

Quantitative distribution and optical purities of the secondary alcohols obtained after fermentation with Penicillium citrinum are very similar to those isolated from coconut or corn (Figure 4). A combination of stereospecific reduction and following enantioselective metabolization may be one of the keys to explain the ratios of enantiomers of aliphatic secondary alcohols observed in natural systems.

Formation of Optically Pure Secondary Alcohols by Yeast Alcohol Dehydrogenase

Alcohol dehydrogenase (ADH) from baker's yeast (Saccharomyces cerevisiae) is a major enzyme involved in the oxidation of secondary alcohols and the reduction of methyl ketones, respectively. The stereochemical course of the oxidation has been investigated using racemic butan-2-ol and octan-2-ol as substrates; only the (S)-enantiomers of these alcohols were converted to the corresponding ketones (22,23). On the other hand the enantioselectivity of yeast ADH in the reduction of ketones has been unclear. When whole yeast cells are employed the reduction of methyl ketones also leads to mainly (S)-configurated alcohols, however the optical purities of these alcohols are only moderate (24). Mac Leod et al. (24) rationalized this lack of 100% stereoselectivity by the assumption that alcohol dehydrogenase may be the only enzyme involved in this reduction but that it is only partially stereoselective when acting on these unnatural substrates.

By means of capillary gas chromatographic determination of the optical purities of formed products we could demonstrate that yeast alcohol dehydrogenase catalyzes not only the oxidation of racemic secondary alcohols but also the reduction of the corresponding methyl ketones in highly stereoselective manner.

Starting from a racemic mixture of an alkan-2-ol only the (S)-enantiomer is converted to the corresponding ketone. A complete oxidation, necessary to eventually obtain the remaining alcohol in optically pure form, was achieved by regeneration of NAD in a coupled reduction of decanal to decanol. Aliphatic methyl ketones were reduced stereoselectively to (S)-alkan-2-ols. NADH, required for this process, was generated from NAD in a coupled oxidation of ethanol to acetaldehyde (Figure 6). The capillary gas chromatographic separations of diastereoisomeric derivatives of a series of aliphatic secondary alcohols obtained by these yeast ADH catalyzed reactions are shown in Figure 7.

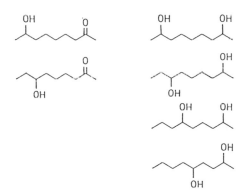

Figure 5. Hydroxylated metabolites isolated after addition of nonan-2-one to a shake culture of Penicillium citrinum.

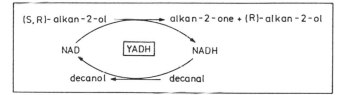

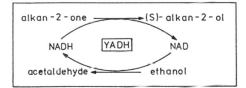

Figure 6. Formation of optically pure secondary alcohols by yeast alcohol dehydrogenase catalyzed oxidation of racemic mixtures of alkan-2-ols and reduction of alkan-2-ones.

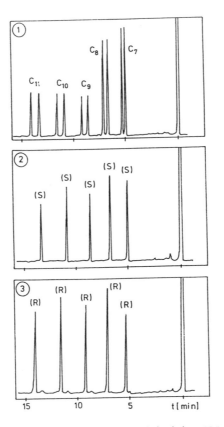

Figure 7. Capillary GC-separation of (R)-(+)-PEIC derivatives of
C₇-C₁₁-alkan-2-ols (DB 210, 30 m x 0.32 mm i.d., df = 0.25 μm;
150-240 °C; 2 °C/ min.).
(1) Racemic alcohols;
(2) (S)-enantiomers obtained by reduction of about 2 mg of alkan-
2-ones in 4 ml Tris-HCl buffer pH 8.1 (0.1 m) containing 0.2 ml
ethanol, 3 mg NAD and 2 mg YADH (specific activity towards ethanol:
280 U/ mg); incubation time: 6 hours;
(3) (R)-enantiomers obtained by oxidation of about 1 mg of racemic
alkan-2-ols in 4 ml buffer pH 8.1 containing 4 mg decanal, 3 mg NAD
and 2 mg YADH; incubation time: 6 hours.

These results demonstrate that whole cells of Saccharomyces cerevisiae must contain oxidoreductases others than alcohol dehydrogenase which are involved in the reduction of methyl ketones and catalyze the formation of (R)-alkan-2-ols.

Outlook

The examples presented in this chapter demonstrate that a combination
of various analytical approaches and the selection of suitable model
systems can add valuable information to our knowledge about pathways
and enzymes involved in the biosynthesis of chiral volatiles. Some of
the techniques need further improvement, e.g. by use of radioactively
labeled precursors the detection threshold of metabolites can be
lowered significantly; addition of precursors in concentrations comparable to those in natural plant or microbial systems would be possible. The investigation of the enantioselectivity of enzymes has to
be emphasized, eventually not only enzymes commercially available or
easily accessible in microorganisms but also those active in plant
systems have to be studied.
   The knowledge about the stereochemical properties of enzymes catalyzing the biosynthesis of chiral volatiles is not only interesting
from a strictly scientific standpoint of view; it is also an essential basis for future improvement of natural flavor and aroma by ge-
netical engineering of plants and microorganisms.

Literature Cited

1. Leitereg, T. G.; Guadagni, D. G.; Harris, J.; Mon, T. R.;
    Teranishi, R. J. Agic. Food Chem. 1971, 19, 785.
2. Mosandl, A.; Heusinger, G.; Gessner, M. J. Agric. Food Chem. 1986,
    34, 119.
3. Mosandl. A.; Deger, W. Z. Lebensm. Unters. Forsch. 1987, 185, 379.
4. Creveling, R. K.; Silverstein, R. M.; Jennings, W. G. J. Food
    Sci. 1968, 33, 284.
5. Näf-Müller, R.; Willhalm, B. Helv. Chim. Acta 1971, 54, 1880.
6. Silverstein, R. M.; Rodin, J. O.; Himel, C. M.; Leeper, R. W.
    J. Food Sci. 1965, 30, 668.
7. Ohta, H.; Kinjo, S.; Osajima, Y. J. Chromatography 1987, 409, 409.
8. Tressl, R.; Engel, K.-H.; Albrecht, W.; Bille-Abdullah, H. In
    Characterization and Measurement of Flavor Compounds; Bills, D. D.;
    Mussinan, C. J., Eds.; ACS Symposium Series No. 289; American
    Chemical Society: Washington, DC, 1985; pp. 43-60.
9. Tressl, R.; Engel, K.-H. In Progress in Flavour Research 1984;
    Adda, J., Ed.; Elsevier Science Publication B. V.: Amsterdam,
    1985; pp. 451-455.
10. Tressl, R.; Engel, K.-H.; Albrecht, W. In Adulteration of Fruit
    Juice Beverages; Nagy, S.; Attaway, J. A.; Rhodes, R. C., Eds.;
    Marcel Dekker, Inc.: New York, 1988; in press.
11. Heidlas, J.; Engel, K.-H.; Tressl, R. Eur. J. Biochem. 1988,
    172, 633.

12. Heidlas, J.; Engel, K.-H.; Tressl, R. manuscript in preparation.
13. Friedmann, E. Die Naturwissenschaften 1931, 450.
14. Lemieux, R. U.; Giguere, J. Canad. J. Chem. 1951, 29, 678.
15. Sih, Ch.; Ching-Shih, C. Angew. Chemie 1984, 96, 556.
16. Tressl, R.; Engel, K.H. In Analysis of Volatiles; Schreier P., Ed.;
    Walter de Gruyter, Berlin, New York 1984; pp. 323-342.
17. Engel, K.-H. In Bioflavour '87; Schreier, P., Ed.; Walter de
    Gruyter: Berlin, 1988; pp. 75-88.
18. Lawrence, R. C.; Hawke, J. C. J. Gen. Microbiol. 1968, 51, 289.
19. Kinderlerer, J. L. Food Microbiol. 1984, 1, 23.
20. Kinderlerer, J. L.; Kellard, B. Phytochemistry 1984, 23, 2847.
21. Kinderlerer, J. L. Phytochemistry, 1987, 26, 1417.
22. Van Eys, J.; Kaplan, N. J. Am. Chem. Soc. 1957, 79, 2782.
23. Dickinson, F. M.; Dalziel, K. Nature 1967, 214, 3.
24. MacLeod, R.; Prosser, H.; Fikentscher, L.; Lanyi, J.; Mosher, H. S.
    Biochemistry 1964, 3, 838.

RECEIVED September 23, 1988

Chapter 3

Aroma Development in Ripening Fruits

P. Dirinck, H. De Pooter, and N. Schamp

Laboratory for Organic Chemistry, Faculty of Agricultural Sciences, State University of Ghent, Coupure Links 653, B–9000 Ghent, Belgium

Instrumental aroma analysis, involving headspace sampling, analysis and identification of aroma compounds by gas chromatography-mass spectrometry allows evaluation of the different flavor determining parameters in fruits. In apples e.g. aroma compounds, mainly esters, are formed gradually during ripening in a manner which parallels the respiration climacteric towards a maximum. Headspace gas chromatography permits following the complete ripening process. Measurement during the early picking period, allows prediction of the earliest acceptable harvest date for storage apples. After storage in controlled atmosphere (CA-storage) the normal ripening pattern is disturbed and ester production diminishes as a function of storage time.
Volatile analysis by non-destructive headspace techniques is also an interesting tool for flavor formation studies e.g. by treatment of intact apples with ester precursors (carboxylic acids, aldehydes, alcohols).

It is generally recognised that in fruit and vegetable production more attention should be given to the hidden sensory quality parameters, such as flavor and texture. These quality attributes are the result of a number of pre- and post-harvest factors and are closely related with fruit ripening. Paillard distinguishes external and internal factors influencing aroma formation in fruits (1). The first ones are associated with the culture of the plant and post harvest treatments, the second are in connection with the metabolic regulation of the fruit. A survey is given in Figure 1.
   It is not possible to cover these different aspects for a variety of fruit products. Therefore we selected to discuss the knowledge of the influencing parameters on apple flavor and to focus on some recent analytical achievements, which are of importance for the study of these effects. Excellent work has already been done concerning the separation and identification of aroma giving compounds. The state of the art up to 1982 on apple flavor has been clearly reviewed by Dimick and Hoskin (2), covering important studies by

0097–6156/89/0388–0023$06.00/0
© 1989 American Chemical Society

EXTERNAL FACTORS

PRE-HARVEST : SOIL/HYDROPONIC CULTURE
              FERTILIZATION
              CLIMATE/IRRIGATION

PICKING DATE-MATURITY
POST-HARVEST TECHNOLOGY : DURATION OF STORAGE
                          STORAGE CONDITIONS :
                          TEMPERATURE
                          HUMIDITY
                          GAS COMPOSITION

INTERNAL FACTORS

GENETIC CONTROL : CULTIVAR
METABOLIC REGULATION : ETHYLENE
                       RESPIRATION

Figure 1 . Factors influencing flavor formation in fruits.

Drawert (3), Flath (4), Guadagni (5), Williams (6) and many others.
Some compounds have been claimed to be important contributors to
apple aroma : e.g. ethyl 2-methylbutanoate, n.hexanal, trans-2-
hexenal and 4-methoxyallylbenzene. However it seems that apple aroma
is not a matter of a limited number of character impact compounds but
is due to complex mixture of alcohols, aldehydes, C_1-C_6 esterified
acids, estragol and terpenes (7). Recently Yajima et al. (8) identified 22 new apple flavor components in the steam distillate of
Kogyoku apples, which correspond to American Jonathan apples. In an
interesting approach using "Charm" analysis on forty different apple
cultivars Cunningham et al. indicated the importance to apple odour
of β-damascenone, butyl-,3-methylbutyl- and hexyl hexanoates, along
with ethyl, propyl and hexyl butanoates.
     Flavor formation in fruit products has also extensively been
reviewed (10). A distinction can be made between the primary aroma
components, which are biosynthesized by the whole fruit and secondary
aroma compounds (e.g. hexanal, 2-hexenal), formed after disruption of
the cells during processing or chewing (11). The importance of the
peel for aroma formation has also been stressed by several authors
(12). An extensive literature on the respiration climacteric (13),
the role of ethylene (14) and the enzymes and substrates required for
biosynthesis is available (15).
     The topic of apple flavor should also be considered in relation
to the modern developments in isolation and preparation technology.
The important influence of the sample preparation techniques on the
composition of biologically derived aromas was recently reviewed by
Parliment (16).
     The goal of this article is to present a procedure for aroma
analysis, consisting of dynamic headspace sampling of intact fruits,
followed by high resolution gas chromatography - mass spectrometry.
This procedure allows evaluation of the influence of cultivar,
picking date and controlled atmosphere storage on aroma development
in apples. The analytical system is also of importance for studying
biogenesis of aroma components in intact fruits.

Analysis Technology: Dynamic Headspace Gas Chromatography

Dynamic headspace sampling on Tenax GC, by trapping volatiles, liberated during disintegration of fruits, has been previously described
by the authors (17,18). This procedure allows the isolation of an
instant aroma and imitates the eating mechanism. Dynamic headspace
sampling also offers the possibility for isolation of volatiles,
released from intact apples, which in the case of apple aroma, allows
following the complete ripening process. A typical apparatus for
isolation of volatiles from intact apples is presented in Figure 2.
The dessicators are continuously flushed with air and at regular
intervals for sampling a Tenax adsorber is attached to the outlet of
the dessicator. For quantitative adsorption high amounts of Tenax GC
are used : 5 g of Tenax GC 60/80 mesh, packed in glass tubes (i.d.
1.6 cm, length 10 cm). Sample recovery can be performed by heat
desorption and collection in a cold trap, which allows a sharp
injection on 0.5 mm i.d. capillary columns. Therefore the gas
chromatographs are modified with a selfconstructed injection system,
consisting of a desorption oven and a thermostated two-position six-
port, high temperature injection valve. The full experimental
details of the analysis system have been described before (19). As
an illustration of the results obtained by this system in Figure 3
the chromatogram of Spartan variety apple volatiles, isolated by
dynamic headspace sampling of the intact fruits, is presented
together with the identified compounds.

Influence of Picking Date

As aroma is one of the key factors in flavor quality, it can be used
as a criterion for evaluation of flavor quality of apples. By following the evolution of the volatile composition a complete picture of
the dynamic flavor quality process can be obtained. In Figure 4 the
sum of esters is presented as a function of days of ripening in standard conditions (18°C) for Golden Delicious apples.
Late picked apples show an immediate aroma development and reach a
considerable higher maximum compared to early picked apples. From
this picture one can conclude that for optimum apple quality the
picking date should be related to the consumption period.
Figure 4 also illustrates the well known fact that too early
picked apples will never develop their full flavor. On the contrary
for apples picked after the optimum harvest date storage ability is
reduced. Establishing the proper time to harvest apple cultivars in
relation to storage disorders and in order to ensure high quality
after long-term CA storage requires special consideration. Post
harvest ripening in apples is associated with color changes, softe-
ning of the fruit flesh, changes of the sour-sweet balance and a rise
of ethylene. Several of these parameters have been used for determination of the optimum harvest date. Unfortunately these harvest in-
dices have no predictive value. As aroma is an exponentially evalua-
ting parameter during the early picking period, we have used the sensitivity of the described analytical procedure for prediction of optimum harvest date for storage apples. At regular intervals in the
premature stage the sum of esters or butyl acetate (for faster
analysis) is measured after 2 days of ripening in standard conditions. The optimum harvest time for storage apples can be predicted

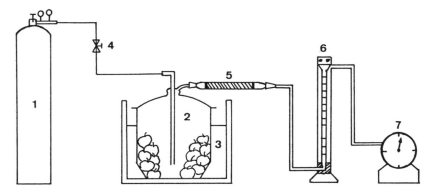

Figure 2.   Apparatus for dynamic headspace sampling of volatiles, released from intact apples.   1 = high purity purge air, 2 = dessicators, 3 = thermostated waterbath (18°C), 4 = fine metering valve, 5 = Tenax adsorber, 6 = flow meter, 7 = wet-testmeter.

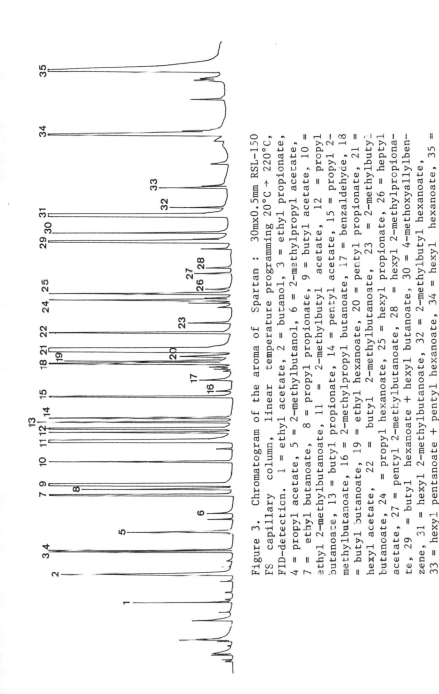

Figure 3. Chromatogram of the aroma of Spartan : 30mx0.5mm RSL-150 FS capillary column, linear temperature programming 20°C → 220°C, FID-detection. 1 = ethyl acetate, 2 = butanol, 3 = ethyl propionate, 4 = propyl acetate, 5 = 2-methylbutanol, 6 = 2-methylpropyl acetate, 7 = ethyl butanoate, 8 = propyl propionate, 9 = butyl acetate, 10 = ethyl 2-methylbutanoate, 11 = 2-methylbutyl acetate, 12 = propyl butanoate, 13 = butyl propionate, 14 = pentyl acetate, 15 = propyl 2-methylbutanoate, 16 = 2-methylpropyl butanoate, 17 = benzaldehyde, 18 = butyl butanoate, 19 = ethyl hexanoate, 20 = pentyl propionate, 21 = hexyl acetate, 22 = butyl 2-methylbutanoate, 23 = 2-methylbutyl butanoate, 24 = propyl hexanoate, 25 = hexyl propionate, 26 = heptyl acetate, 27 = pentyl 2-methylbutanoate, 28 = hexyl 2-methylpropiona- te, 29 = butyl hexanoate + hexyl butanoate, 30 = 4-methoxyallylben- zene, 31 = hexyl 2-methylbutanoate, 32 = 2-methylbutyl hexanoate, 33 = hexyl pentanoate + pentyl hexanoate, 34 = hexyl hexanoate, 35 = α-farnesene.

by linear regression between the logarithm of the butyl acetate concentration and the picking date.   The results for Golden Delicious
apples in the season 1986 are presented in Figure 5.   As evaluated by
storage experiments and subsequent aroma analyses optimum harvest
corresponds to a butyl acetate concentration of 0.4 µg/kg apples/6 1
dynamic headspace.

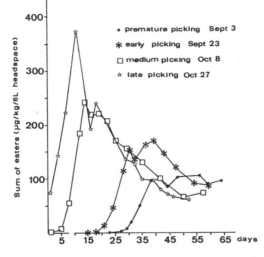

Figure 4.   Influence of date of gathering on the evolution of the sum
of Golden Delicious esters as a function of days of ripening in
standard conditions.

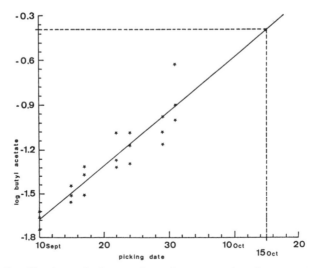

Figure 5.   Prediction of the optimum harvest time for storage apples
by linear regression between log butyl acetate (after 2 days of
ripening) and picking date.

Postharvest Preservation Technology

In most countries an important part of the apple production is stored. Early studies showed that high carbon dioxide levels and low oxygen atmospheres could delay ripening. Choosing the storage circumstances seems largely to have been a question of trial and error and an impressive amount of work has been done about storage in high carbon dioxide and/or low oxygen atmospheres (20), under hypobaric or low ethylene conditions (21,22).

Aroma analysis can be used as an objective criterion for measurement of the evolution of flavor quality as a function of controlled-atmosphere storage time. Some years ago we measured the evolution of aroma development in Golden Delicious apples after removal from CA-storage for six periods during the storage period by means of a headspace technique, which isolated the volatiles released during maceration of the fruits (23). In a latter experiment we used the non-destructive headspace sampling technique with intact apples. Because the same fruits are sampled throughout the complete ripening experiment this technique has several advantages, such as convenience, faster analysis and better reproducibility. An illustration of the aroma development after removal from ventilated controlled atmosphere (Temperature : 0,5°C, 16% carbon dioxide, 5% oxygen) is shown in figure 6. Results objectively indicate an important decrease in aroma quality after long CA-storage.

Postharvest preservation technology is still in progress and more and more attention is given to low oxygen and to low carbon dioxide levels (ULO-storage) and to ethylene scrubbing. Our future research in this field will be directed to the evaluation of these recent developments in order to maintain flavor quality after long storage. The proposed technique can be used together with panel evaluations to compare different storage conditions.

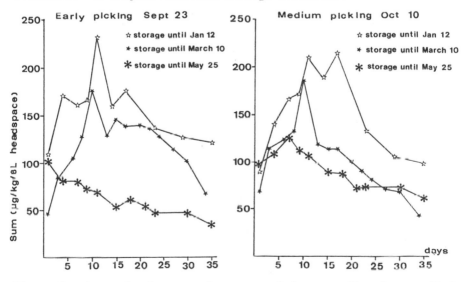

Figure 6. Aroma development after removal from ventilated controlled atmosphere. Influence of storage time and picking date.

Genetic Control : Cultivars

One of the important strategies of the recent quality policy of the
apple industry consist in diversification of the number of cultivars,
which are offered to the consumer. For instance the Belgian production, which around 1980 consisted of 75% Golden Delicious has been
diversified to about 33% Golden Delicious, 24% Jonagold, 18% Boscoop,
6.5% Cox's Orange Pippin, 4.3% Gloster and others. As apple cultivars can differ markedly in aroma character, the decision which cultivars should be grown is of extreme economic importance. Classification of apple cultivars according to their aroma composition can be
helpful for cultivar selection and for detection of relationships
between cultivars.   Information on the aroma patterns of different
apple cultivars is scarce (24,25).

In a recent project we have determined the aroma development of
25 apple cultivars by means of dynamic headspace isolation of the
volatiles released by intact fruits, followed by high resolution gas
chromatography and mass spectrometric identification. Data acquisi-
tion resulted in 25 tables with quantitative data for about forty
compounds at 5 to 8 ripening stages. For classification purposes the
composition near the maximum of aroma development was used and 12
parameters were selected : hexanol, butyl acetate, hexyl acetate,
butyl propionate, hexyl propionate, ethyl butanoate, propyl butanoate, butyl 2-methylbutanoate, hexyl 2-methylbutanoate, hexyl hexanoate, 4-methoxyallylbenzene and α-farnesene. Parameter selection was
guided by visual comparison of the chromatogram and also some analytical and biochemical insights were taken in consideration. Hexanol
was selected as a representative of alcohols, which are characteristic of some apple cultivars. The different esterified C_1 to C_6 acids
were represented by the high concentration butyl and hexyl esters,
except for the butanoates. As respectively butyl butanoate/ethyl
hexanoate and hexyl butanoate/butyl hexanoate were not well separated
on the RSL-150 (SE-52) capillary column, we used ethyl and propyl
butanoate for representing the butanoates and disregarded butyl
hexanoate. Furthermore 4-methoxyallyl-benzene, responsible for the
spicy note in apples, and α-farnesene as the main sesquiterpene were
used for classification.

Data processing was performed by Principal Component Analysis on
a personal IBM computer.   In a first set the procentual values of
the selected parameters of the 25 apple cultivars were used. A
projection of the 12-dimensional into a 2-dimensional space is
presented in Figure 7. The reduced space presents in an optimal way
(75.5% of the original variance) the relations between the objects.
In the same plane the relations between the variables are given and
can be related to the objects. Figure 7 shows that a large group of
cultivars is characterised by high concentration of acetates. High
concentrations of hexanol and ethyl butanoate are typical for another
group, composed of Nico, Granny Smith, Summerred and Paulared.
Boscoop and Jacques Lebel are outliers and are characterised respectively by high α-farnesene and high hexyl 2-methylbutanoate.

In a second data set the relationships between the closely
related "acetate"-type cultivars was examined. The results for 17
cultivars are presented in a 3-dimensional space (88% of the original
variance) in Figure 8. From this picture it is clear that f.i.
Jonagold (Golden Delicious x Jonathan) en Golden Delicious have very

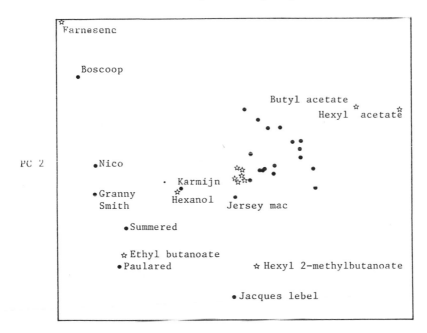

PC 1

Figure 7. Principal component analysis of 25 apple cultivars.

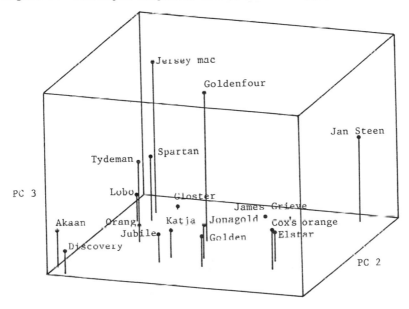

PC 1

Figure 8. Principal component analysis of 17 "ester"-type apple cultivars.

similar aroma patterns. The results are especially of importance for
cultivars with future potential for the apple industry, e.g. Elstar
(Golden Delicious x Ingrid Marie), which has a similar aroma pattern
as Cox's Orange Pippin.

Metabolic Control : Flavour Formation

Quantitative differences in volatile ester composition of different
apple varieties may be explained by their biogenesis. In apples, the
generation of aroma esters takes place mainly in the peel, is oxygen
dependent and requires the organisation of intact tissue (5). In
incorporation experiments apple disks were treated with carboxylic
acids or aldehydes to study ester formation (26). As we thought that
wounding of the tissues by preparing disks might disturb the normal
behavior of fruits by dearranging the tissue organisation, intact
apples were treated with aldehydes and carboxylic acids in the vapor
phase (27,28), and volatiles were collected by dynamic headspace
sampling on Tenax GC. Subsequent analysis gave a much more detailed
picture of the aroma development compared to the direct headspace
sampling in incorporation experiments (29,30). As an illustration of
the effect of adding aldehydes, Figure 9 shows a graphical picture of
the volatiles, isolated by dynamic headspace sampling 1 day after
treatment of Golden Delicious apples with pentanal and hexanal.
Treatment of intact fruits with carboxylic acids or aldehydes led to
the enhanced formation of volatile esters and showed that the fruits
possessed activity for esterification, for α- and β-oxidation, for
aldehyde reduction, and also were capable of reducing carboxylic
acids into alcohols probably by the way of the corresponding
aldehydes (31,32).

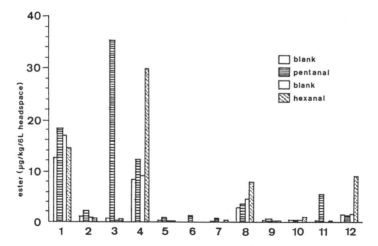

Figure 9.   Comparison of ester formation by Golden Delicious apples
one day after treatment with pentanal and hexanal. 1 = butyl acetate, 2 = 3-methylbutyl acetate, 3 = pentyl acetate, 4 = hexyl acetate,
5 = butyl propionate, 6 = pentyl propionate, 7 = hexyl propionate,
8 = butyl butanoate, 9 = butyl 2-methylbutanoate, 10 = hexyl 2-
methylbutanoate, 11 = hexyl pentanoate, 12 = hexyl hexanoate.

Conclusions

Aroma development in ripening apples is dependent on a number of external and internal factors. Dynamic headspace sampling of the volatiles from intact fruits, followed by high resolution gas chromatography or g.c.-m.s. anlysis is a convenient procedure for the study of these factors.

Acknowledgment

The "Instituut tot Aanmoediging van het Wetenschappelijk Onderzoek in Nijverheid en Landbouw" is thanked for supporting this investigation.

Literature cited

1. Paillard, N. In Flavour '81; Schreier, P., Ed.; Walter de Gruyter : Berlin, New York, 1981, pp. 479-507.
2. Dimick, P.S.; Hoskin, J.C. Crit. Rev. Food Sci. Nutr., 1983, 18, 387-408.
3. Drawert, F.; Heimann, W.; Emberger, R.; Tressl, R. Chromatographia, 1969, 2, 57.
4. Flath, R.A.; Black, D.R.; Guadagni, D.G.; McFadden, W.H.; Schultz, T.H. J. Agric. Food Chem., 1967, 15, 29.
5. Guadagni, D.G.; Bomben, J.L.; Harris, J.G. J. Sci. Food Agric., 1971, 22, 115.
6. Williams, A.A.; Tucknott, O.G.; Lewis, M. J. Sci. Food Agric. 1977, 28, 185.
7. Van Straten, S.; de Vrijer, F.L.; de Beauveser, J.C.; Eds. Volatile Components in Food, 4th ed., 1977, Central Institute for Nutrition and Food Research T.N.O. Zeist, The Netherlands.
8. Yajima, I.; Yamai, T.; Nakamura, M.; Sakakibara, N.; Hayashi, K. Agric. Biol. Chem. 1984, 48, 849.
9. Cunningham, D.G.; Acree, T.E.; Bamard, J.; Butts, R.M.; Braell, P.A. Food Chem. 1986, 19, 137.
10. Salunkhe, D.K.; Do, J.Y. Crit. Rev. Food Sci. Nutr. 1976, 8, 161.
11. Drawert, F.; Kuchendauer, F.; Brickner, H.; Schreier, P. Chem. Mikrobiol. Technol. Lebensm. 1976, 5, 27.
12. Guadagni, D.G.; Bomben, J.L.; Hudson, J.S. J. Sci. Food Agric. 1971, 22, 110.
13. Rhodes, M.J.C. In The Biochemistry of Fruits and their Products; Hulme, A.C., Ed.; Academic Press : New York, 1970; Vol. 1, p. 521.
14. Yang, S.F. Hort Science 1985, 20, 41.
15. Yabumoto, K.; Yamaguchi, M.; Jennings, W.G. Chem. Mikrobiol. Techn. Lebensm. 1977, 5, 53.
16. Parliment, T.H. In Biogeneration of Aromas; Parliment, T.H., Croteau, R., Eds.; American Chemical Society : Washington, DC, 1986; pp. 34-52.
17. Dirinck, P.; Schreyen, L.; Schamp, N. J. Agric. Food Chem. 1977, 25, 759.
18. Schamp, N.; Dirinck, P. In Chemistry of foods and beverages : recent developments; Charalambous, G., Inglett. G., Eds.; Academic Press : New York, 1982, pp. 25-47.

19. Dirinck, P.; De Pooter, H.; Willaert, G.; Schamp, N. In Analysis of Volatiles; Schreier, P., Ed.; Walter de Gruyter & Co. : Berlin, New York, 1984; pp. 381-399.
20. Kader, A.A. Food Techn. 1980, 34, 51.
21. Burg, S.P.; Burg, E.A. Science 1966, 153, 314.
22. Knee, M. In Ethylene and Plant Development; Roberts, J.A.; Tuckner, G.A., Eds.; Butterwords : London; pp. 297-315.
23. Willaert, G.H.; Dirinck, P.J.; De Pooter, H.L.; Schamp, N.M. J. Agric. Food Chem. 1983, 31, 809.
24. Drawert, F.; Heimann, W.; Emberger, R.; Tressl, R. Z. Lebensmitteluntersuch. u. -Forsch., 1969, 140, 65.
25. Paillard, N.; Lebensm. -Wiss. u. Technol., 1975, 8, 34.
26. Paillard, N. Phytochemistry 1979, 18, 1165.
27. De Pooter, H.L.; Dirinck, P.J.; Willaert, G.A.; Schamp, N.M. Phytochemistry 1981, 20, 2135.
28. De Pooter, H.L.; Montens, J.P.; Willaert, G.A.; Dirinck, P.J.; Schamp, N.M. J. Agric. Food Chem., 1983, 31, 813.
29. Ambid, C.; Fallot, J. Bull. Soc. Chim. Fr., 1980, 104.
30. Knee, M.; Hatfield, S.G.S. J. Sci. Food Agric., 1981, 32, 593.
31. De Pooter, H.L.; D'Ydewalle, Y.E.; Willaert, G.A.; Dirinck, P.J.; Schamp, N.M. Phytochemistry, 1984, 23, 23.
32. De Pooter, H.L.; Van Acker, M.R.; Schamp, N.M. Phytochemistry, 1987, 26, 89.

RECEIVED September 23, 1988

Chapter 4

Nonvolatile Conjugates of Secondary Metabolites as Precursors of Varietal Grape Flavor Components

Patrick J. Williams, Mark A. Sefton, and Bevan Wilson

The Australian Wine Research Institute, Private Mail Bag, Glen Osmond, SA 5064, Australia

An examination of hydrolysates produced by glycosidase enzyme or pH 3.2 acid treatment of C_{18} reversed-phase isolates from juices of "non-floral" Vitis vinifera vars. Chardonnay, Sauvignon Blanc and Semillon demonstrated that these grapes contain conjugated forms of monoterpenes, C_{13} norisoprenoids, and shikimic acid-derived metabolites. The volatile compounds obtained hydrolytically from the conjugates were produced in sufficient concentration to permit ready analysis by GC/MS. The products of pH 3.2 hydrolysis have sensory significance when assessed in a neutral wine. The study further develops the precursor analysis approach as a technique to facilitate research into varietally specific constituents of grapes.

The most extensively studied flavor compounds of Vitis vinifera grapes are monoterpenes (1). These compounds, along with other well recognized constituents of grape berries such as anthocyanins, hydroxycinnamate esters and tannins, are categorized by plant biochemists as secondary metabolites. One of the defining characteristics of secondary metabolites is their propensity to accumulate in particular organs or tissues of the host (2). This accumulation occurs presumably either as a result of products being formed in terminating biosynthetic pathways, or because of relatively slow product turnover. However, in the case of grape monoterpene flavor compounds, as with the flavor constituents of most other fruits, experience proves that accumulation is not one of their significant properties. Flavor compounds of grapes are trace constituents only, and even in berries of the most floral varieties, i.e. the Muscats, monoterpene flavor compounds are present at concentrations of only 1-2 mg/L of juice (3).

Research over the last few years has revealed that the flavor compounds are not end products of monoterpene biosynthesis in the grape. Oxidative pathways, leading to flavorless polyhydroxylated forms of the monoterpenes, are active in Vitis vinifera. Also, the

0097-6156/89/0388-0035$06.00/0
© 1989 American Chemical Society

grape's monoterpenes, in both unoxidized forms and as polyols, occur
as glycosidic conjugates in the fruit (1). More recent evidence (4)
indicates that glycosylation takes place after the extra hydroxyl
groups are introduced into the monoterpene skeleton, which is
consistent with the view that glycosylation is usually the terminal
step of any biosynthetic pathway (5). In keeping with this proposal
analyses have shown that, for most monoterpene dependent varieties,
the glycosides accumulate in the berries to a greater extent than the
free compounds (3, 4).

A second property of many of the monoterpene glycosides is their
high reactivity in weakly acid media (1). Acid-catalyzed hydrolysis
at pH 3.0-3.5 of these non-volatile and flavorless glycosidic
compounds, can give volatile monoterpenes, some of which have
significant sensory properties. Accordingly, the presence of many
monoterpenes found in grapes and wines can be accounted for by
non-enzymatic reactions (1).

Thus, two diametrically opposite consequences follow from these
secondary transformations of the grape's monoterpenes. Conjugation
of monoterpenes by glycosylation is a process diminishing the pool of
free compounds and thereby denudes the fruit of flavor.
Simultaneously, however, some of the conjugates are activated by
glycosylation and can thus regenerate volatiles, different from the
aglycons, by non-enzymatic mechanisms.

Recognition of these phenomena in grapes and wines (1), and the
importance therefore of the glycosides as precursors of flavor in
those systems (6-8) has stimulated much research interest in the role
of non-volatile flavor precursors in other fruits (9-11), processed
fruits (12), and leaf products (13, 14). For convenience this line
of investigation is described here as the "precursor analysis
approach" to flavor research.

This paper discusses new developments in this field of research
and in particular the application of the precursor analysis approach
to the study of flavor compounds of non-monoterpene containing grape
varieties.

Glycosylation and the Flavor Properties of Non-Floral Grape Varieties

From observations of the monoterpene aglycons enzymatically released
from the glycosidic fractions of different floral varieties, it was
concluded that glycosylation was a pathway by which the grape "fixes"
the monoterpene alcohol composition (15). Thus, analysis of aglycons
can furnish a metabolic fingerprint for the monoterpenes of a
particular variety, indicating inter alia, the extent of accumulation
of individual compounds as glycosides, the degree of hydroxylation of
the terpenes, and the type of products formed by oxidation. This
interpretation has, up until now, been assumed to be limited to that
relatively small number of floral grapes which are dependent on
monoterpenes for their varietal flavor (6).

In non-monoterpene dependent grape varieties the flavor is often
delicate and subtle, and knowledge of the chemical composition of the
flavor compounds is almost non-existent. A notable exception in this
regard is the recent confirmation of the role of alkyl
methoxypyrazines in contributing to the varietal character of
Sauvignon Blanc wines (16, 17). However, for other grapes in this

non-monoterpene dependent category, which incidentally make up the majority of the world's winemaking varieties, even basic information on the types of compounds which might be contributing to the characteristic flavors is not available.

The major reason for this deficiency is the tiny quantity of volatiles present in juices of these non-floral grapes. Thus, in spite of constant advances being made in the techniques of GC/MS for the identification of flavor compounds (18) there has been little progress in this area of varietal grape flavor research. Accordingly, it was clear that an alternative approach to the problem should be sought.

To this end a strategy was developed based on the hypothesis that similar metabolic processes operate on the unknown flavor compounds of non-floral varieties to those which transform the monoterpenes of the floral grapes, i.e. oxidative hydroxylation of some compounds, followed by conversion of these, and of unoxidized volatiles, to flavorless glycosides. Additionally, it was assumed that glycosylation would 'fix' the composition of the alcohols and related derivatives and allow them to accumulate in a manner similar to that observed for the monoterpene aglycons. Accordingly, analysis of the glycoside fractions of non-floral grapes, i.e. precursor analysis, could lead not only to a rationalization of those volatiles formed by non-enzymatic steps in these fruits, but also to an elucidation of the free impact flavor compounds from which the conjugates were formed.

The Precursor Fractions of Sauvignon Blanc, Semillon and Chardonnay Grapes

The C_{18} reversed-phase (RP) liquid chromatographic procedure used for isolation of monoterpene glycosides from floral varieties (19), was applied to juices of three non-floral grapes, i.e. Sauvignon Blanc, Semillon and Chardonnay. Hydrolysis of the isolated fractions, both enzymatically and by aqueous acid at pH 3.2, gave a range of volatiles (see Table I) confirming that these grapes also accumulate low molecular weight compounds as conjugated derivatives. Acid hydrolyses were carried out by heating anaerobically the C_{18} RP isolates from 4.5L of juice at $50°$ for 1 month in 90mL of aqueous saturated potassium hydrogen tartrate, adjusted to pH 3.2 with tartaric acid.

Sensory Significance of the Precursor Fractions and Their Hydrolysates

Before progressing further with the chemical analysis of the hydrolysates, the sensory significance of these and of the C_{18} RP precursor materials was assessed.

Samples of the unhydrolysed C_{18} RP concentrates, the enzyme released aglycons, and the pH 3.2 acid hydrolysates from each grape variety, were presented to a panel for duo-trio aroma assessment in a neutral wine medium. Untreated fractions and aglycons were presented for evaluation at concentrations equivalent to twice that in the original juices, and the acid hydrolysates were presented at 1.5 times their original juice concentrations. Other conditions used for

Table I. Aglycons and pH 3.2 Hydrolysis Products from C_{18} RP Isolates

Compound[a]	Sauv. Blanc		Chardonnay		Semillon		Evidence[b]	Ref
	Agly	H+	Agly	H+	Agly	H+		
MONOTERPENES								
Furan linalool oxide 1 **1**		++		+++		++	A	1
Furan linalool oxide 2 **1**		++		++		++	A	1
Hotrienol **2**		+				++	A	1
Nerol oxide **3**		+				+	A	1
α-Terpineol **4**	++						A	1
A pyran linalool oxide **5**			+	++		+	A	1
2,6-Dimethylocta-3,7-diene-2,6-diol **6**	++	++	+	++		++	A	1
Geraniol **7**	++					++	A	1
2,6-Dimethyloct-7-ene-2,6-diol **8**		+++		++		+++	A	1
cis-1,8-Terpin **9**		+++		++		+++	A	1
trans-1,8-Terpin **10**		+++		++		++	A	1
(Z)-2,6-Dimethylocta-2,7-diene-1,6-diol **11**	++		+			++	A	4
(E)-2,6-Dimethylocta-2,7-diene-1,6-diol **12**	+++		++			+++	A	4
Unknown paramenthenediol 1		++					B	
Unknown paramenthenediol 2		+		+			B	21
Unknown paramenthenediol 3	++++	+			++	++	B	21
NORISOPRENOIDS								
6-Methylhept-5-en-2-one	++						B	22
2,6,6,-Trimethylcyclohex-2-ene-1,4-dione		++		++		++	C	23
Vitispiranes **13**		++		++		++	A	24
2,2,6-Trimethylcyclohexane-1,4-dione				+		+	A	25
1,1,6-Trimethyl-1,2-dihydronaphthalene **14**		+		++		+	A	19
Damascenone **15**		++		++		++	C	19
Actinidol 1 **16**		++		+++		+++	A	26
Actinidol 2 **16**		+++		+++		+++	A	26

Table I. Continued

Compound[a]	Sauv. Blanc		Chardonnay		Semillon		Evidence[b]	Ref
	Agly	H+	Agly	H+	Agly	H+		
Actinidol 3 **17**				+		++	B	26
2-(3-Hydroxybut-1-enyl)-2,6,6-trimethylcyclohex-3-en-1-ones **18**		+++		+++		+++	A	27
3,4-Dihydro-3-oxo-actinidol **19**				++			A	25
3-Oxo-β-damascone **20**	+++		++				E	21
3-Hydroxy-β-damascone **21**	++	+	+	++	+	++	C	21
3,9-Dihydroxymegastigm-5-en-7-yne **22**	++		++		+		E	28
3-Oxo-α-damascone **23**					+		E	21
3-Oxo-α-ionol **24**[c]	+++	+	++	+++	+++	++	C	25
3-Oxo α-ionone **25**			+			+	C	25
9-Hydroxymegastigma-4,6-dien-3-one (isomer 1) **26**				++			E	29
Blumenol C **27**	+++		+		+		E	30
9-Hydroxymegastigma-4,6-dien-3-one (isomer 2) **26**				++			E	29
Unknown A (isomer 1)[d]	++				++		B	21
Unknown A[e] (isomer 2)[d]	+++	++	++	+++	+		B	21
Unknown B	+++		+++		+++		B	21
Vomifoliol **28**	+++	++	+++	+++	+++	+++	C	25
Dehydrovomifoliol **29**	++	+		++	+	++	C	25
SHIKIMATE-DERIVED								
Benzaldehyde		++		++		++	C	
Benzyl alcohol	++++	+++	++	++	+++	+++	A	31
2-Phenylethanol	+++	+++	++	+++	++	+++	A	31
2-Hydroxybenzoic acid methyl ester				+		+	B	
Benzoic acid		++		+++	+	++	C	
Phenylacetic acid				++		++	C	
2,3-Dihydroxybenzoic acid methyl ester	+						C	
Vanillin **30**	+	++		+++		++	C	21
Methyl vanillate **31**			+	+	+	+	C	21
Acetovanillone **32**	+			++	+	++	D	22
4-Hydroxybenzaldehyde		+++		+++		+	C	
Tyrosol **33**	++		+		++	++	D	22
Benzophenone		++					D	22
4-Hydroxybenzoic acid methyl ester				++			D	22
4-Hydroxyacetophenone		++		++		+	C	

Continued on next page.

Table I. Continued

Compound[a]	Sauv. Blanc Agly	Sauv. Blanc H^+	Chardonnay Agly	Chardonnay H^+	Semillon Agly	Semillon H^+	Evidence[b]	Ref
2,5-Dihydroxybenzoic acid methyl ester	+++	+	+	+	+		C	
Vanillic acid 34				++			D	22
Raspberry ketone 35					++	++	D	22
Zingerone 36					+		A	21
Dihydroconiferyl alcohol 37			+	+	+		A	21
Benzyl benzoate	+					+	B	
Syringaldehyde 38	+			+			D	22
4-Hydroxybenzoic acid				+++			D	22
Methyl syringate 39				++			B	
Acetosyringone 40				+			D	22
Coniferyl alcohol 41			+		+		A	21
Syringic acid 42				++			D	22

Constituents in each class are listed in order of increasing retention time on a J&W DB 1701 fused silica capillary G.C. column. Quantities were estimated by a comparison of the total ion count for the mass spectrum of each peak with that of an n-octanol internal standard; +, 0.1-1 µg/L; ++, 1-10 µ g/L; +++, $\overline{1}$0-100 µ g/L; ++++, 100-500 µg/L. [a] For compound structures see Figure 1. [b] A, previously identified in this laboratory; B, tentatively identified from interpretation of the mass spectrum; C, symmetrical peak enhancement and identical mass spectrum on coinjection with an authentic sample; D, comparison of mass spectrum with published spectrum; E, comparison of mass spectrum and retention time with that of material available commercially or synthesized in the laboratory. [c] Both diastereoisomers observed in each of the acid hydrolysates, only one in each of the aglycon fractions. [d] This is unknown 17 of Ref. 21. [e] This is unknown 16 of Ref. 21.

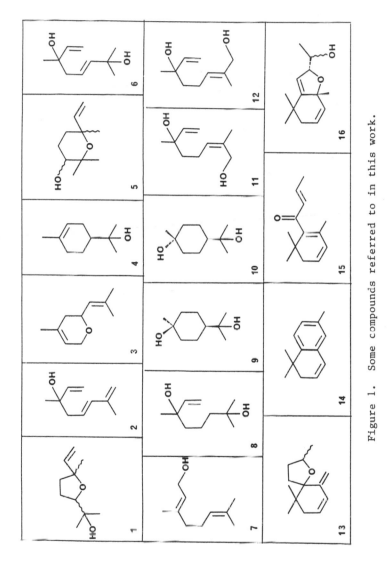

Figure 1. Some compounds referred to in this work.

Continued on next page.

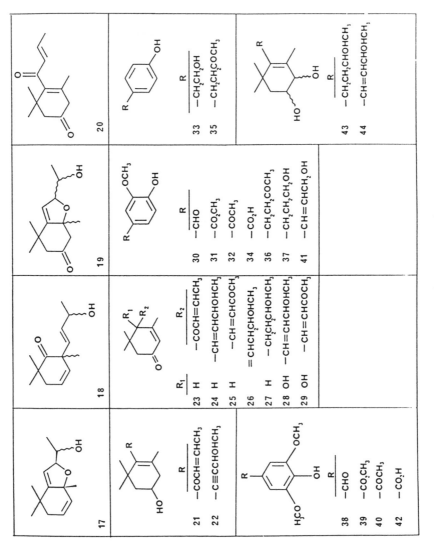

Figure 1. Continued.

the sensory studies have been described previously (20). Results of these experiments are given in Table II.

Table II. Summary of Sensory Panel Responses for Duo-Trio Aroma Evaluations of Glycosidase Enzyme Hydrolysates, pH 3.2 Acid Hydrolysates, and Untreated C_{18} RP Fractions From Three Non-Floral Grape Varieties

	Significance of Responses		
Variety	Untreated[a]	Enz. Hydrol.[a]	Acid Hydrol.[b]
Sauvignon Blanc	NS	*	**
Chardonnay	NS	NS	***
Semillon	NS	NS	***

NS = not significant, * = significant at 5%, ** = significant at 1%, *** = significant at 0.1%.
a, n = 13 judges; b, n = 16 judges

It is clear from the data in Table II that the acid hydrolysates of the precursor fractions had a highly significant effect on the aroma of the wines to which they were added. In contrast, neither the original precursor materials nor the glycosidase-released aglycons, with the possible exception of those from Sauvignon Blanc, could be detected by the panel.

These observations are in accord with the results of a previous study which demonstrated that the aglycons from a Chardonnay glycosidic fraction were not a significant source of added aroma to the wine from which it had been obtained (20). The present work emphasises the importance of mild acid hydrolysis and the associated chemical pathways to yield flavor significant volatiles in these non-floral varieties.

Sensory descriptive analyses of the acid hydrolysates are in progress, and preliminary data from these experiments for the products given by the Chardonnay precursors at pH 3.2 indicate that oak wood and vanilla descriptors are important to the sample of this variety.

Chemical Composition of the Precursor Fraction Hydrolysates

It can be seen from the data in Table I that the aglycons and the products given by mild acid treatment of the C_{18} RP retained material from the three grape varieties were made up of monoterpenes, thirteen carbon and other norisoprenoids, together with a group of shikimate-derived aromatic compounds. It is also apparent that the acid hydrolysates were more complex than the respective aglycon fractions, with the acid-derived products containing many compounds not seen in the material released by the glycosidase enzyme.

In addition to the products listed in Table I, a large number of, as yet, unidentified compounds was found, mainly in the acid hydrolysates. Of this group, five were monoterpenes, twenty-one were norisoprenoid and twenty-three were in the aromatic or phenolic class. The unknown norisoprenoids, aromatic and phenolic compounds, were observed predominantly in the Chardonnay and Semillon hydrolysates. At least six nitrogen–containing compounds and two γ-lactones were also observed, together with a large number of minor constituents for which no structural category has yet been assigned.

The Monoterpenes. Monoterpenes are not major components of non-floral grapes ($\underline{1}$) although some of these compounds were present in each of the three varieties. In a previous survey of monoterpene diols in floral and non-floral grapes, ($\underline{Z}$)- and ($\underline{E}$)-2,6-dimethylocta-2,7–diene–1,6–diols 11 and $\overline{12}$ were not found in the Chardonnay samples examined ($\underline{4}$). Observation of these diols in the Chardonnay here, and reports of their occurrence in several other Chardonnay clones ($\underline{7}$), indicates that these compounds 11 and 12 are common monoterpenes of grapes. No products obviously derivable from a conjugate of diols 11 and 12 were seen in the acid hydrolysates, which is consistent with the previously reported stability of the 1,6–diol system in these compounds ($\underline{4}$).

The 1,8–terpins 9 and 10, which are monoterpenes of high flavor threshold, were observed in the acid hydrolysates of each variety. These diols are derivable as thermodynamically stable end-products of acid catalyzed rearrangement of monoterpenes which are at the oxidation state of geraniol. However, few compounds at this oxidation state were seen among the aglycons, and the precursors of the terpins 9 and 10 in these varieties have yet to be discovered.

The Norisoprenoid Compounds. The C_{13}-norisoprenoid compounds in Table I can be grouped into four different oxidation levels. Compounds exemplifying these four levels are vitispirane 13, damascenone 15, vomifoliol 28, and dehydrovomifoliol 29. Such subclassification should help in establishing the nexus between hydrolysis product and precursor. Nevertheless, many of the compounds given by acid hydrolysis cannot yet be related directly back to the aglycons observed. Thus, for example, previous research has shown that megastigmatriols 43 and 44, which occur naturally in grapes, give on hydrolysis several of the norisoprenoids observed in this study, i.e. vitispirane 13 from 43, and the actinidols 16 and 17 together with hydrocarbon 14 and the rearranged ketone 18 from 44 ($\underline{27}$). However, neither triols 43 and 44, nor any obviously related isomers, were observed among the aglycons, and precursors of volatiles 13, 14, 16, 17 and 18 have yet to be found in these non-floral varieties.

Conversely, many of the major norisoprenoid aglycons recorded in Table I do not appear to contribute significantly to the products given by acid hydrolysis. The hydrolytic chemistry of vomifoliol 28 and 3–oxo–α–ionol 24 has been studied ($\underline{25}$); only trace amounts of 3–oxo–α–ionone 25 and an isomer of 3,4–dihydro–3–oxoactinidol 19, which are derivable from the former aglycon 28, were observed, and none of the products reported ($\underline{25}$) from 3–oxo–α–ionol 24 were found.

Obviously further work is needed in linking the norisoprenoid components of the precursor fractions to volatiles in the hydrolysates. An important aspect of this will be the elucidation of the transposition mechanism of the oxygen function from position

9 to 7 in the megastigmane skeleton, with the formation of damascenone 15 and relatives 20 and 23.

The Shikimate-Derived Group.   Examination of the aromatic compounds seen here and previously (21) indicate that C_6-C_1, $-C_2$, $-C_3$ and $-C_4$ compounds, showing a variety of hydroxy- and methoxy-substitution patterns, are present in the grape precursor fractions. These products are known to be derived in plants from phenylpropanoids via side-chain degradation and elongation reactions (32, 33). Because of the obvious sensory significance of many of these aromatic compounds, e.g. vanillin 30, raspberry ketone 35 and zingerone 36, the incidence of occurrence, and the origins of individual constituents within this class of secondary metabolite, together with the nature of the conjugating moieties involved with them, are now recognized as subjects of high importance to grape flavor.

Compositional Differences Among the Varieties

Differences among the aglycons and acid hydrolysates of these three non-floral varieties would not be expected to be as obvious as between those of floral and non-floral grapes (15). Nevertheless, this first exploratory investigation into the hydrolytically produced volatiles from the C_{18} RP precursor fractions of Chardonnay, Semillon and Sauvignon Blanc allows differences to be discerned. These are best assessed by considering each of the three groups of compounds individually.
    From Table I it can be seen that Chardonnay contained fewer and lower concentrations of monoterpenes than either the Semillon or Sauvignon Blanc juices. With regard to the norisoprenoid compounds, the Sauvignon Blanc juice appeared to contain a higher concentration of these constituents in glycosidically bound form than did the Chardonnay or Semillon juices. In contrast the C_{18} RP fractions from the latter varieties yielded norisoprenoids, including the unknowns in this group, in greater abundance and variety following pH 3.2 hydrolysis than did the Sauvignon Blanc sample. More phenolic and aromatic constituents were yielded by the Chardonnay precursor fraction than from the other two varieties. Close examination of the phenolic and aromatic group of compounds suggests further distinguishing features. For example, while C_6-C_1 and C_6-C_2 compounds were common to all of the grapes, Chardonnay and Semillon appeared to produce greater amounts of aromatic compounds with longer side-chains than did the Sauvignon Blanc. The aromatic substitution patterns may also be influenced by varietal genetics. Thus guaiacyl-substituted compounds were observed mostly in Chardonnay and Semillon, while the more heavily substituted syringyl constituents occurred almost exclusively in the Chardonnay hydrolysates.

Conclusion

Previous work on precursor fractions from grapes had indicated that analysis of the conjugated constituents would facilitate research in the varietal specific constituents of fruit (21). This study further develops that research and proves that conjugated forms of

many secondary metabolites are present in the C_{18} RP fractions from
non-floral grapes and mild acid hydrolysis of these gave a number
of volatiles, many of which are of known flavor importance.
Furthermore, the acid hydrolysates from the three varieties were
demonstrated to have sensory significance in wine. The research
also indicates that differences in composition among the acid- and
glycosidase enzyme hydrolysates of the three juice samples will
allow varietal differences to be observed.

Future work must investigate the conjugating moieties involved
with the grape secondary metabolites. Plants are known to employ
sugars with both ester and glycosidic linkages, as well as several
non-sugar moieties for conjugation purposes (34). Thus considering
glycosides alone may not give a complete picture of the bound flavor
compounds of the fruit. This is emphasised by the data in Table I
in which many of the acid hydrolysis products cannot be rationalized
in terms of the aglycons released by Rohapect C from the precursor
fractions.

It will also be necessary to relate the structures of the
conjugates in the precursor fractions with the composition of the
hydrolysates. This is particularly important where polyfunctional
molecules are involved allowing more than one site of conjugation,
because acid hydrolysis of one conjugate of a particular secondary
metabolite may yield different products from that of an
alternatively or polyconjugated species.

Finally, and most importantly, the acid hydrolysates and enzyme
deconjugated products can be used to facilitate investigations into
those trace volatiles which are free in the grape or finished wine,
and which are responsible for the varietal flavor.

The precursor analysis approach should now be seen as a useful
complement to traditional methods of flavor analysis of fruits. The
latter methods are often limited to the painstaking processes of
isolation and identification of those trace constituents which are
directly responsible for flavor. The precursor analysis approach
takes advantage of the evidence provided by Nature when secondary
metabolites, including flavor compounds, are biochemically
transformed and accumulated by the fruit.

Acknowledgments

We thank G. Gramp & Sons Pty Ltd, Thomas Hardy & Sons Pty Ltd, and
S. Smith & Son Pty Ltd for generously donating samples of grape
juices and wines. The Grape and Wine Research Council is thanked
for supporting this work.

Literature Cited

1. Strauss, C.R.; Wilson, B.; Gooley, P.R.; Williams, P.J.
    In Biogeneration of Aromas; Parliment, T.H.; Croteau, R., Eds.;
    ACS Symposium Series No. 317; American Chemical Society:
    Washington, DC, 1986; pp222-242.
2. Haslam, E. Nat. Prod. Rep. 1986, 3, 217-249.
3. Gunata, Y.Z.; Bayonove, C.L.; Baumes, R.L.; Cordonnier, R.E. J.
    Chromatogr. 1985, 331, 83-90.

4. Strauss, C.R.; Wilson, B.; Williams, P.J. J. Agric. Food Chem.
   1988, 36, 569-573.
5. Hösel, W. In The Biochemistry of Plants. Secondary Plant
   Products; Conn, E.E., Ed.; Academic: London, 1981; Vol. 7, p725.
6. Williams, P.J.; Strauss, C.R.; Aryan, A.P.; Wilson, B.
   Proc. 6th Aust. Wine Industry Technical Conference, 1986, p111.
7. Versini, G.; Scienza, A.; Dalla Serra, A.; Dell'Eva, M.; Rapp,
   A. In Bioflavour '87; Schreier, P., Ed.; de Gruyter:Berlin, New
   York, 1988.
8. Braell, P.A.; Acree, T.E.; Butts, R.M.; Zhou, P.G. In
   Biogeneration of Aromas; Parliament, T.H.; Croteau, R. Eds.; ACS
   Symposium Series No. 317; American Chemical Society: Washington,
   DC, 1986; pp75-84.
9. Winterhalter, P.; Schreier, P. In Bioflavour '87; Schreier, P.,
   Ed.; de Gruyter:Berlin, New York, 1988.
10. Engel, K-H.; Tressl, R. J. Agric. Food Chem. 1983, 31, 998-1002.
11. Heidlas, J.; Lehr, M.; Idstein, H.; Schreier, P. J. Agric. Food
    Chem. 1984, 32, 1020-1021.
12. Chen, C-C.; Kuo, M-C.; Liu, S-E.; Wu, C-M. J. Agric. Food Chem.
    1986, 34, 140-144.
13. Nitz, S.; Fischer, N.; Drawert, F. Chem Mikrobiol. Technol.
    Lebensm. 1985, 9, 87-94.
14. Fischer, N.; Nitz, S.; Drawert, F. Z.Lebensm. Unters. Forsch.
    1987, 185, 195-201.
15. Strauss, C.R.; Wilson, B.; Williams, P.J. Proc. 6th Aust. Wine
    Industry Technical Conference, 1986, p117.
16. Harris, R.L.N.; Lacey, M.J.; Brown, W.V.; Allen, M.S. Vitis
    1987, 26, 201-207.
17. Lacey, M.J.; Brown, W.V.; Allen, M.S.; Harris, R.L.N. Proc. 2nd
    Int. Symp. Cool Climate Viticulture and Oenology, 1988, p344.
18. Cronin, D.A.; Caplan, P.J. In Applications of Mass Spectrometry
    in Food Science; Gilbert, J., Ed.; Elsevier: London, 1987;
    Chapter 1, 1.
19. Williams, P.J.; Strauss, C.R.; Wilson, B.; Massy-Westropp, R.A.
    J. Chromatogr. 1982, 235, 471-480.
20. Noble, A.C.; Strauss, C.R.; Williams, P.J.; Wilson, B.
    Proc. 5th Weurman Flavor Res. Symp., 1987, p383.
21. Strauss, C.R.; Gooley, P.R.; Wilson, B.; Williams, P.J.
    J. Agric. Food Chem. 1987, 35, 519-524.
22. Heller, S.R.; Milne, G.W.A. EPA/NIH Mass Spectral Data Base,
    U.S. Department of Commerce and The National Bureau of
    Standards: Washington, DC, 1978; Vol. 1-4.
23. Fujimori, T.; Kasuga, R.; Matsushita, H.; Kaneko, H.; Noguchi,
    M. Agric. Biol. Chem. 1976, 40, 303-315.
24. Simpson, R.F.; Strauss, C.R.; Williams, P.J. Chem. Ind.
    (London), 1977, 663-664.
25. Strauss, C.R.; Wilson, B.; Williams, P.J. Phytochemistry 1987,
    26, 1995-1997.
26. Dimitriadis, E.; Strauss, C.R.; Wilson, B.; Williams, P.J.
    Phytochemistry 1985, 24, 767-770.
27. Strauss, C.R.; Dimitriadis, E.; Wilson, B.; Williams, P.J.
    J. Agric. Food Chem. 1986, 34, 145-149.

28. Loeber, D.E.; Russell, S.W.; Toube, T.P., Weedon, B.C.L.;
    Diment, J. J. Chem Soc(C) 1971, 404-408.
29. Lloyd, R.A.; Miller, C.W.; Roberts, D.L.; Giles, J.A.;
    Dickerson, J.P.; Nelson, N.H.; Rix, C.E.; Ayers, P.H.
    Tob. Sci. 1976, 20, 40-48.
30. Aasen, A.J.; Hlubucek, J.R.; Enzell, C.R. Acta. Chem. Scand.
    1974, B28, 285-288.
31. Williams, P.J.; Strauss, C.R.; Wilson, B.; Massy-Westropp, R.A.
    Phytochemistry 1983, 22, 2039-2041.
32. Gross, G.G. In The Biochemistry of Plants. Secondary Plant
    Products; Conn, E.E., Ed.; Academic:London, 1981; Vol. 7, p301.
33. Barz, W.; Köster, J.; Weltring, K-M.; Strack, D. Ann. Proc.
    Phytochem. Soc. Eur. Vol. 25, 1985, p307.
34. Barz, W.; Köster, J. In The Biochemistry of Plants. Secondary
    Plant Products; Conn, E.E., Ed.; Academic: London, 1981;
    Vol. 7, p35.

RECEIVED August 5, 1988

Chapter 5

Sotolon

Identification, Formation, and Effect on Flavor

Akio Kobayashi

Ochanomizu University, Laboratory of Food Chemistry, 2–1–1, Ohtsuka, Bunkyo-ku, Tokyo, Japan

Sotolon, 4,5-dimethyl-3-hydroxy-2(5H)-furanone, was isolated and identified as a flavor impact compound from raw cane sugar and has a very low threshold value. The aroma characteristic changes from caramel-like at low concentrations to curry-like aroma at high concentrations. For this reason the compound, which had already been synthesized and also found in its natural form, was not associated with the characteristic brown lump sugar aroma. The syntheses of sotolon homologues and their enantiomers provided some information about the structure-aroma relationship. The formation of sotolon was confirmed in a model system composed of glutamic acid and pyruvate, the latter being estimated as a reaction product of an amino-carbonyl reaction. As a flavor impact compound, sotolon was found in botrytized wine and roasted tobacco. The presence of sotolon in these products is indispensable for a flavor with high sensory qualities.

Cane sugar is one of the oldest agricultural products known to man and originated in the tropics. Despite the primitive technology for sugar processing that was used in the early stage of its development, raw cane sugar became edible and even palatable because of its acceptable flavor. Thus cane sugar history contrasts with the development and manufacture of beet sugar produced in northern Europe since the 19th century.

Brown lump sugar is prepared from calcified cane juice simply by boiling down into a solid form, which is widely used as an ingredient of traditional Japanese cakes in addition to other partially refined raw sugars. These are not only used as sweeteners but also as flavoring ingredients. The characteristic caramel-like, burnt-sweet aroma of raw cane sugar is known to be formed at the last stage of the heating and condensing process for sugar cane juice, and the flavor impact compound (FIC) has been estimated to be a nonenzymatic product formed during the browning reaction. Since

0097–6156/89/0388–0049$06.00/0

Takei et al.(1) first reported in 1936 the separation and
identification of a volatile component in raw cane sugar and
molasses, many researchers have tried to identify the FIC of the
characteristic sugary aroma in cane juice or raw cane sugar
products. Furfural, hydroxymethyl furfural, maltol, isomaltol and
3-methyl-2-hydroxy-2-cyclopenten-1-one (cyclotene) have been
identified as typical amino-carbonyl reaction products with a burnt
sugary aroma; however, their threshold values are too high to
explain the strong characteristic aroma of raw cane sugar. Through
many research studies it has become clear that this sugary aroma
could be concentrated in a specific fraction, which implies that one
or a few FICs must be present, although the amount of FICs has been
too little to identify by the usual analytical method.

Separation and Identification

As the content of the aroma substance was estimated to be extremely
low, we used cane molasses as the starting material for our study,
as the sugary aroma was already condensed in it and the material
could be supplied in bulk by the manufacturer. One ton of cane
molasses was first extracted with acetone, and after evaporating the
solvent, the low molecular weight organic materials were extracted
continuously with ether. The extract was then divided into basic,
acidic, weakly acidic and neutral fractions; by an organoleptic
evaluation the sugary aroma appeared strongly in both the weakly
acidic and neutral fractions. As the yield of the neutral fraction
was much higher than that of the other fractions, the combined
weakly acidic and neutral fractions were further fractionated by
silica gel column chromatography. The aroma was concentrated in
fraction 11, which was then separated by packed-column GC with peak
sniffing ("nasal appraisal"). At a retention time of about 60 min,
the strong sugary aroma was noted, although many peaks overlapped in
this zone. In the next step, the effluent between t_R 55 and 65 min
was trapped repeatedly and analysed by high-resolution gas chromatography (HRGC) combined with mass spectrometry (MS). This separation
scheme is summarized in Figure 1, (2,3) and the resolution of the GC
trapping fraction by HRGC is shown in Figure 2.

By GC-MS analysis, peaks 36, 37 and 39 were estimated to be
3-hydroxy-4,5-dimethyl-2(5H)-furanone, acetate of hydroxymethyl-
furfural and 4-pentyl-2-pentenolide, respectively. At this stage,
the sample was too small to apply other analytical methods;
therefore, we tried to synthesize all the possible compounds using
the synthetic approaches described in the following section. None of
thethree synthetic products showed the characteristic sugary aroma
that we had recognized in each separated fraction; however, the
yield of fraction 11-GC TRAP from molasses was calculated to be ca.
1 ppm, and the concentration of FIC in molasses was estimated to be
in the order of ppm or ppb from its peak area in the whole gas
chromatogram. By diluting these synthetic products in water to the
concentration of 1.0 ppm, 3-hydroxy-4,5-dimethyl-2(5H)-furanone

Cane Molasses _____ 1000 kg

    Acetone (1400g), stir overnight, separate upper layer
    Distill off acetone

Oleoresin _____ 31.6 kg

    Continuous ether extraction for 40hr
    Distill off ether

Ether Extract _____ 752 g

    Separate basic and strongly acidic fractions

Weakly Acidic and Neutral Fractions _____ 90.2 g

    Silica gel column chromatography

Fraction 11 _____ 17.3 g

    Preparative gas chromatography

Fr. 11-GC TRAP (3.3% of whole peak area) _____

HRGC-MS

Figure 1. Scheme for Extraction and Fractionation

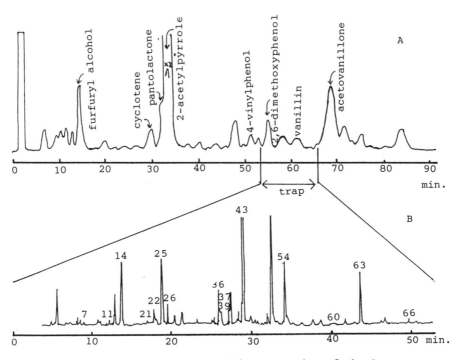

Figure 2. Gas Chromatographic Separation of the Aroma
Components in Cane Molasses
(A) Preparative Gas Chromatography of Fr.11.
(B) Gas Chromatogram of Fr.11-GC TRAP (peak numbers in
    B correspond to those in Table 1.)

showed a strong burnt sugary aroma reminiscent of those in the separated fractions from molasses, all the gas-chromatographic and spectrometric data for the synthesized product coinciding with those for the natural one. Therefore, it became clear that a high concentration of the synthetic product produced another effect on the olfactory organ. In the next step to confirm the contribution to the sugary aroma in molasses, all the compounds identified in fraction 11-GC TRAP were reconstructed with the concentrations appearing on gas chromatogram B in Fig. 2, as is shown in Table 1,(A) without and (B) with sotolon.

An organoleptic omission test on this mixture confirmed 3-hydroxy-4, 5-dimethyl-2(5H)-furanone to be the most important FIC for a raw cane sugar aroma, and the other main components in Table 1, i.e., vanillin, maltol, 4-pentyl butanolide, 4-vinyl phenol and 2,6-dimethoxy phenol, seemed to improve the overall sugary aroma. From these results, we(4) gave the trivial name "sotolon" to 2-hydroxy-4,5-dimethyl-2(5H)-furanone, which is built up from "soto" (raw sugar in Japanese) and "olon" (enol lactone) as a main FIC in raw cane sugar.

Table 1. The synthetic mixtures of the identified compounds in Fr. 11-GC trap

Peak No.	Compound	A Conc. (%)	B Conc. (%)
7	Furfuryl alcohol	0.20	0.20
11	Damascenone	0.13	0.12
14	Guaiacol	0.72	0.69
21	5,6-Epoxy-β-ionone	0.04	0.04
21	Maltol	0.21	0.20
22	2-Acetylpyrrole	2.53	2.45
25	Phenol	5.70	5.51
26	4-Pentylbutanolide	3.38	3.27
36	Sotolon	-	3.27
37	5-Acetoxymethylfurfural	0.84	0.81
39	4-Pentyl-2-pentenolide	0.17	0.16
43	2,6-Dimethoxyphenol	67.12	64.92
50	Isoeugenol	0.17	0.17
54	4-Vinylphenol	1.90	1.84
60	5-Hydroxymethylfurfural	1.27	1.22
63	Vanillin	14.14	13.68
66	Acetovanillone	1.48	1.42

Synthetic Approaches

The synthesis of sotolon can be traced back to 1947(5). Sulser et al.(6) subsequently improved the synthetic method and obtained sotolon as a homolog of 3,4-dialkyl-2-hydroxy-butenolactone. They also identified II in a vegetable-protein hydrolysate as a flavoring compound(6), although they described the aroma character of I and II as Maggiherb-like at a concentration of 5-1 ppm, and as a walnut-like herbal aroma at 0.1 ppm with threshold values between 1-0.5 ppb. On the other hand, Rödel and Hempel(7) described the aroma character of I as fruity and alcoholic at the concentration of 100 ppm in water, and as a Maggi herbal and celery-like aroma at 50 ppm. We suggest the following reasons for there having been no description of a sugary aroma for synthetic I (sotolon): (a) The previous studies concerned the flavor substances of a vegetable protein hydrolysate, and when synthetic products showed a herbal aroma, they did not attempt an organoleptic test at lower concentrations. (b) These unsaturated lactones are unstable and easily polymerized to a viscous oil, although this change was often overlooked because of the strong odor. We, therefore, reinvestigated the synthesis of several sotolon homologs by following the general synthetic route described in Figure 3, and evaluated the aroma characters in their pure states at a concentration close to the respective threshold value. The structure and purity of distilled sotolon (bp 0.2 mm Hg 84-86°C) and the other homologs were confirmed by GC, MS, IR, PMR and CMR. The threshold values evaluated by an experienced panel from Takasago Perfumery Co. Ltd. are summarized in the bottom line of Figure 3 (unpublished data). In an earlier report (4), we assigned the threshold value of synthetic sotolon to be 0.002 ppb, although the new results have reduced this value to 10^{-3} ppb, and those of some longer alkyl-substituted homologues are ever lower. Sotolon is unstable in its pure state even in a refrigerator, although it can be preserved as a diluted solution in high polar solvents such as water, ethylene glycol or glycerin. These results suggest that sotolon is easily polymerized to a higher molecular weight product and thus gradually loses its strong aroma character.

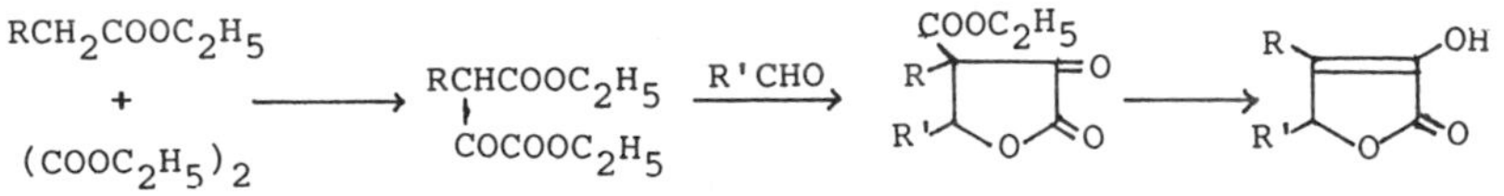

	I	II	III	IV	V	VI	VII
R	Me	Me	Me	Me	Me	Et	Et
R'	Me	Et	Pro	Bu	iso-Bu	Me	Et
Threshold Value (ppb)	1×10^{-3}	1×10^{-5}	1×10^{-5}	5×10^{-6}	1×10^{-3}	2.5×10^{-4}	2.5×10^{-5}

Figure 3.   General Synthetic Route to 3,4-Dialkyl-2-hydroxy-butenolactones and their Threshold Values

All the compounds produced a burnt sugary aroma, which became more burnt and heavy as the substituted alkyl chain increased in length. It is interesting that the ethyl substituted lactone II has a 100 times lower threshold value than that of sotolon, and that this compound has been considered to be an FIC in the protein hydrolysate. This was because it had been prepared from threonine by heating with hydrochloric acid and subsequent dehydration, hydrolysis, condensation (Aldol type) and decarboxylation, and it showed a strong curry-like or herbal aroma at concentrations higher than 1 ppm.

Sotolon has an asymmetric carbon in its molecule, and therefore, stereospecific syntheses(8) of sotolon enantiomers would be effective for correlating the stereostructure and olfactory sensation. If the aroma character turned out to be quantitatively or qualitatively different between these enantiomers, we could expect to elucidate whether sotolon is a naturally occurring compound or a product formed by chemical reaction during the sugar manufacturing process. Starting from D-and L-tartaric acids, the respective 2,3-epoxybutenes, (2R,3R) and (2S,3S), were prepared by the known method. 1,3-Dithiane-2-carboxylic acid condensed with the respective epoxides and removal of the thioacetal group gave optically active sotolones. The $[\alpha]_D^{23.5}$ value in ether for (R)-sotolon was -6.5° and for (S)-Sotolon was +7.1°. The synthetic scheme is summarized in Figure 4.

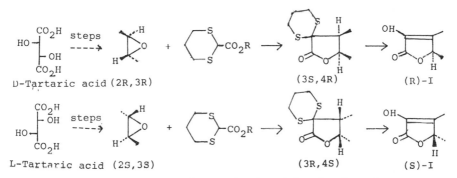

D-Tartaric acid (2R,3R)   (3S,4R)   (R)-I

L-Tartaric acid (2S,3S)   (3R,4S)   II (S)-I

Figure 4.   Stereospecific Synthesis of Sotolon Enantiomers

Both the enantiomers showed the same aroma character at the same concentration near the threshold value of (±) sotolon; moreover, there was no difference in the insect attractancy (house fly and cockroach) among the two enantiomers and the racemate of sotolon. Later, we(9) tried to analyze the aroma compound in fresh sugar cane juice and could not identify sotolon in the same fraction as that separated from cane molasses. These results suggest that sotolon was present in a racemic form prepared by mutual interaction of the constituents in sugar cane juice.

Formation of Sotolon

A burnt, sugary aroma is representative of some cooked foods. A number of sugary aroma compounds are listed below:

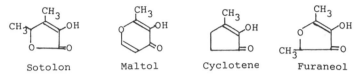

Sotolon   Maltol   Cyclotene   Furaneol

Figure 5.   Sugary Aroma Compounds Having a Common Partial Structure

These have a partial common structure, i.e. enolyzed α-diketone and α-hydrogen to the enolizable carbonyl is substituted by a methyl group. For a characteristic sugary aroma, the presence of enol-hydrogen was essential, because the acetyl ester or methyl ether of sotolon showed no characteristic aroma. The extraordinarily low threshold value of sotolon may be due to its coexistence with hydrophilic and hydrophobic (alkyl-substituted lactone) partial structures in one molecule. The lower threshold value of the ethyl-substituted homolog (II) than that of sotolon may also be explained from the balance of these opposite physicochemical properties.

Apart from sotolon, the other compounds in Fig. 5 can be explained as the products of a Maillard reaction, and their carbon skeletons simply originate from the active Amadori intermediate; in other words, they still preserve the straight carbon chain structure of monosaccharides. In spite of being a simple C_6 lactone, sotolon has a branched carbon skeleton, which implies another formation process in the Maillard reaction. Sulser et al.(6) reported that ethyl sotolon (II) was prepared from threonine with sulfuric acid, and that 2-oxobutyric acid, a degradation product of threonine, was a better starting material to obtain II. This final reaction is a Claisen type of condensation, which would proceed more smoothly under alkaline conditions. As we(10) obtained II from 2-oxobutyric acid (see figure 6) with a high yield in the presence of potassium carbonate in ethanol, a mixed condensation of 2-oxobutyric and 2-oxo-propanoic (pyruvic) acids was attempted under the same conditions, and a mixture of sotolon (22% yield) and II were obtained; however, the

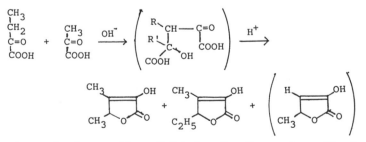

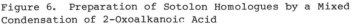

Figure 6.   Preparation of Sotolon Homologues by a Mixed Condensation of 2-Oxoalkanoic Acid

self-condensation product of 2-oxopropanoic acid, desmethylsotolone, was not identified in the reaction mixture.
In the actual manufacturing process for sugar, there exist various Maillard reaction series, and pyruvate seemed to be a degradation product of carbohydrate. On the other hand, 2-oxobutyric acid could be derived from glutamic acid which is the main free amino acid in sugar cane juice as well as aspartic acid.(11) An equimolar amount of glutamic and pyruvic acids was dissolved in water, the pH was adjusted to 8 with potassium carbonate and the mixture boiled for 4 hours. Sotolon was identified in the ether extract of the reaction mixture by GC-MS the yield was below 0.1 %. From this data, the glutamate seemed to have been oxidized to α-keto glutarate in the presence of pyruvate, before it was condensed with the pyruvate and subsequent decarboxylation to yield sotolon. In practice, the characteristic aroma of brown lump sugar appears at the last manufacturing stage when the high-sugar-content liquid (Bx. 73) is heated at 135 °C; therefore, sotolon must be formed at this stage following the complex amino-carbonyl reaction just described. A speculative formation mechanism for sotolon and its homologues is shown in Fig. 7, in which an unstable keto acid oxidizes an amino acid to the α-keto acid, which then condenses with another keto acid to form the furanone structure. The last stage of the reaction is decarboxylation, which would be dependent on the pH and temperature conditions.

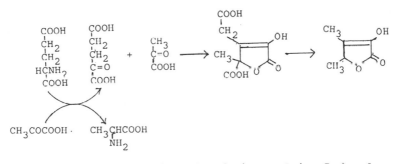

Figure 7.   Formation of Sotolon during an Amino-Carbonyl Reaction

Flavoring Roles of Sotolon

When Japanese rice wine (sake) is kept under unsuitable conditions, it develops an off-flavor with a burnt or soy-sauce (shoyu)-like odor. Takahashi et al.(12) identified the main component as sotolon before our identification, and they claimed that the sotolon concentration was between 140-430 ppb in aged sake. This concentration is much higher than the threshold value of sotolon, and its aroma character would change from sugary to herbal or curry-like. The high sotolon content in aged sake could be a reason for the off-flavor defect. The formation of sotolon in aged

sake has also been proposed by the same author. Contrary to the
heating conditions during cane sugar manufacture, the formation of
this off-flavor proceeds at room temperature and under acidic
conditions. They claimed that 2-oxobutyric acid and acetaldehyde
reacted to produce sotolon in the presence of many organic acids. In
this case, 2-oxobutyric acid was thought to have been derived from
threonine (2.7 mM in Japanese sake). On the other hand, sotolon was
found as the most important flavoring material in botrytized
wine, (13), which is highly praised for its noble sweet flavor formed
by infection with Botrytis cinerea fungus in ripe grapes. The subtle
amount of flavoring material was separated by DEAE-sephadex column
chromatography and identified by GC-MS. A quantitative analysis was
also performed by mass fragmentography, and the concentration of
sotolon in this wine was calculated to be 5-20 ppb. The same content
levels were found in botrytized wines from Japan, France and
Germany; however, in normal wines, the levels were lower than 1 ppb.
By an organoleptic omission test on the aroma constituents in
botrytized wine, it became clear that sotolon plays an important
role in this special wine character. These results obtained from
aged sake and botrytized wine show that sotolon can develop the off-
flavor or be a desirable flavor substance depending on its
concentration in the beverage. Sotolon has also been found in
volatiles from roasted tobacco (14), and the addition of the
condensate of the volatiles to cigarettes improved their aroma
properties and decreased the offensive odor and taste (15). By
combining the fractionation of the volatile condensate and an
organoleptic evaluation, sotolon proved to be one of the active
compounds with 4-hydroxy-2,5-dimethyl-3(2H)-furanone (furaneol).
Therefore, the formation of sotolon is possible under a wide range
of conditions, from mild fermentation and preservation to more
intensive roasting conditions. As described above sotolon shows its
burnt sugary aroma at a ppm or ppb level; therefore, in order to
prove the role of sotolon in the sugary aroma products, it became
necessary to analyze sotolon quantitatively and qualitatively by
applying the effective concentration for sensitive micro-analysis.
By following this analytical system, we attempted to verify the
presence of sotolon and establish its contribution to the aroma of
raw cane sugar (not published). The aroma fraction was concentrated
by silica gel column chromatography as already described and
analyzed by GC-MS with a SIM (selected ion monitoring) system, in
which the selected ion on the mass spectrometer was fixed and each
GC peak scanned for this fragment ion. By fixing ions, the
sensitivity of MS can be greatly increased to the pg order, and this
makes it possible to identify quantitatively a very low
concentration component such as sotolon. We selected the molecular
ion (M^+ 128) for the SIM analysis and deduced the amount of sotolon
in imported raw sugar (a raw material for the sugar refining
process) to be between 0.1-0.01 ppb, which is enough to contribute
the characteristic aroma to raw cane sugar.

Conclusion

With the development of various analytical instruments and
techniques, it has become possible to identify minor but important
aroma constituents. As the volatile aroma compounds generally have a

simple chemical structure, a FIC with very low concentrations in food was identified as a known compound after much tiresome and time-consuming work, even when using sophisticated methods of separation and identification. Our finding of sotolon as an FIC in raw cane sugar is one such example; moreover, through this study, we began to recognize that some aroma compounds change their odor character with concentration. If such a compound has a very low odor threshold value, it is important that an olfactory evaluation of the odor character in a diluted state near this threshold value be made. To investigate such an aroma compound, it became necessary to integrate all the results obtained by the various scientific approaches of olganoleptic evaluation, syntheses of a target compound or its homologues, chemical or biochemical transformation in a model system and modern instrumental analyses. Flavor chemistry in the future will be of concern to an ever-widening scientific field, and will mature into a more sophisticated science.

Literature Cited
 1. Takei, S.; Imaki, T.   Bull. Inst. Phys. Res. Tokyo 1936, 15, 124.
 2. Abe, E.; Nakatani, Y.; Yamanishi, T.; Muraki, S.   Proc. Japan Acad. 1978, 54B, 542.
 3. Tokitomo, Y.; Kobayashi, A.; Yamanishi, T.; Muraki. S.   Proc. Japan Acad. 1980, 56B, 452.
 4. Tokitomo, Y.; Kobayashi, A.; Yamanishi, T.; Muraki, S.   Proc. Japan Acad. 1980, 56B, 457.
 5. Schinz, H; Hinder, M.   Helv. Chim. Acta 1947, 39, 1347.
 6. Sulser, H.; Habeggar, M.; Buchi, M.   Z. Lebens-Unters. Fortsch. 1972, 148, 215.
 7. Rodel, W.; Hempel, U.   Die Nahrung 1974, 18, 133.
 8. Okada, K.; Kobayashi, A.; Mori, K.   Agric. Biol. Chem. 1983, 47, 1071.
 9. Tokitomo, Y.; Kobayashi, A.; Yamanishi, T.   Agric. Biol. Chem. 1984, 48, 2869.
10. Nose, M.; Kobayashi, A.; Yamanishi, T.; Matsui, M.; Takei, S.   Nippon Nogeikagaku Kaishi 1983, 57, 557.
11. Meade, G. P.; Chen. J. C. P.   Sugar and Sugar Canes; Wiley International: New York, 1977; p.27.
12. Takahashi, K.; Tadenuma, M.; Sato, S.   Agric. Biol. Chem. 1976, 40, 325.
13. Masuda, M.; Okawa, E.; Nishimura, K.; Yunome, H.   Agric. Biol. Chem. 1984, 48, 2720.
14. Matsukura, M.; Takahashi, K.; Kawamoto, M.; Ishiguro, S.; Matsushita, H.   Agric. Biol. Chem. 1985, 49, 3335.
15. Matsukura, M.; Ishiguro, S.   Agric. Biol. Chem. 1986, 50 3101.

RECEIVED November 11, 1988

Chapter 6

Role of Oxidative Processes in the Formation and Stability of Fish Flavors

C. Karahadian and R. C. Lindsay

Department of Food Science, University of Wisconsin–Madison, Madison, WI 53706

The development of characterizing fish aromas and flavors involve both enzymic and nonenzymic oxidative reactions. Lipoxygenase-derived carbonyls and alcohols contribute the distinctive planty-green aroma notes to fresh fish that vary with species with regards to compounds and concentrations. Additionally, key secondary oxidation products contribute distinctive aromas characteristic to certain fish species. In salmon, co-oxidation of polyunsaturated fatty acids of fish oils with salmon-specific carotenoid pigments leads to the formation of a characterizing cooked salmon flavor compound, and changes the ratio of carbonyl compounds formed compared to that for pure fish oil. Nonenzymic autoxidation reactions which are predominant in highly unsaturated fish lipids, can be directed by the use of tocopherol-type antioxidants to manipulate oxidized flavors and thus influence the quality of fish aromas and flavors. Secondary oxidation of aldehydes to acids by peracids is responsible for the formation of short chain fatty acids (C_4 to C_8). These acids do not appear to contribute characterizing flavors and aromas in oxidizing fish lipids.

The development of both desirable and undesirable fishy flavors has long-been a concern to the seafood and fishery industry (1-6). Oxidative processes occurring through enzymic and nonenzymic mechanisms initiate hydroperoxide formation in fish lipid systems that are responsible for the formation of the short chain carbonyls and alcohols which exhibit distinct fish-like flavors and aromas. Because the generation of fresh fish aroma compounds involves some of the same polyunsaturated fatty acid precursors and oxidative pathways as autoxidation, it has been a tedious task to differentiate the mechanisms and aroma compounds

0097–6156/89/0388–0060$06.00/0

contributing desirable fresh characteristics from those causing
deteriorated fishy aromas and flavors.

Initiation processes for hydroperoxide siting on
polyunsaturated fatty acids in fish occur by mechanisms
encountered broadly in lipid oxidation, and include enzymic
hydroperoxidation (7-11), singlet oxygen addition to double bonds
(12-17), and classical Farmer-type H-abstraction and molecular
oxygen addition to the pentadiene free radical (18-22).
Considerable attention has been given to the identification of
compounds derived from enzymically (23-27) as well as
autoxidatively-produced (28-32) hydroperoxides of long chain n-3
fatty acids that are responsible for character-impact fish-like
flavors and aromas. It now has been well documented that C_6,
C_8 and C_9 aldehydes generated enzymically from n-3 fatty acids
(23-27) are responsible for much of the planty, green-type aromas
encountered in very fresh fish and seafoods.

Autoxidative reactions in fish or fish oils can also lead to
green flavors (32), but these flavors are accompanied by elevated
concentrations of compounds contributing general oxidized flavors
as well as the 2,4,7-decatrienals and c-4-heptenal which are
primarily responsible for burnt, oxidized-fishy or cod liver
oil-like flavors.

Enzymically generated fish flavors and aromas

Lipoxygenase (10-11, 23-27), NADH-dependent microsomal oxidase
(33-36), and myeloperoxidase (37-38) that are endogenous to fish
tissue all apparently can participate in hydroperoxide formation
on unsaturated fatty acids of fish lipids. Of these enzymes,
however, only lipoxygenases provide positional hydroperoxide site
specificity on the long-chain n-3 fatty acids, and such siting
leads to specific compounds that initially cause characteristic
fresh fish aromas. Differences in lipoxygenase enzymes in
freshwater and saltwater species provide distinct profiles of
volatile compounds that account for the species specific flavor
and odor variations (23-27). Generally, both freshwater and
saltwater fish contain substantial levels of hexanal and
unsaturated C_8 alcohols which are responsible for the heavy,
green character notes (23-27).

Nine-carbon compounds, such as 2-nonenal, 2,6-nonadienal and
3,6-nonadienal, occur mainly in freshwater species, and contribute
to the fresh green melon-like quality (23,26,27). Pacific salmon
entering freshwater from saltwater enroute to spawning produce the
nine-carbon compounds, but only the C_6 and C_8 carbonyls and
alcohols are found in saltwater-dwelling salmon (39). Such a
transition to the production of the nine-carbon compounds along
with the C_6 and C_8 compounds has been suggested to be related
to the induction of physiological processes that increase slime
formation, and possibly provide improved osmoregulation by salmon
in freshwater environments. The resulting variations in
lipid-derived compounds during the life cycle of salmon, however,
influence the aroma and flavor of fish harvested at various times.

A 12-lipoxygenase found in trout gill extracts (10-11) is
responsible for the siting of the n-9 hydroperoxide on

eicosapentanoic acid (EPA) and docosahexaneoic acid (DHA). This site specific hydroperoxide formation in conjunction with lyase activity results in the formation of 3,6-nonadienal, 2,6-nonadienal and 3,6-nonadien-1-ol which provides the fresh, green-cucumber quality of certain very fresh fish, such as the lake whitefish (Coregonus clupeaformis; 23,27). The presence of the 15- and 12-lipoxygenases in fish provides specificity for the siting of n-6 and n-10 hydroperoxides on n-3 fatty acids, respectively. These enzymes are postulated to account for the biogenesis of the unsaturated C_6 and C_8 carbonyls and alcohols (39). These aroma compounds contribute a heavy, plant-like character to fresh fish, and their formation parallels similar reactions on shorter-chain, unsaturated fatty acids in fruits and vegetables (40-42). The existence of a 5-lipoxygenase in fish has been suggested (39) in relation to the formation of short chain oxoacids, but evidence for this enzyme is circumstantial at the present time.

The myeloperoxidase/halide system found in fish has been shown to produce elevated levels of superoxide anions and hydrogen peroxide which in turn randomly peroxidizes the olefin bonds of fish lipids (37-38). The NADH-dependent oxidase system, which was isolated as a membrane fraction from fish skeletal muscle, requires ADP and iron ions to initiate hydroperoxidation, but its mechanism of activity is still unclear (33-36).

Singlet oxygen-generated hydroperoxides

Singlet oxygen-mediated hydroperoxidation reactions are random processes of initiation because of the non-selective interactions of singlet oxygen for any olefin in unsaturated systems. The increased energy state of molecular oxygen provided by spin inversion to form the higher energy singlet state allows for direct olefinic interaction with singlet oxygen (16, 17). The mechanism of excitation involves photosensitizers that allow spin conversion to occur, and natural photosensitizers in foods include porphyrin ring structures, such as heme and chlorophyll, which possess conjugated sites of unsaturation for electron delocalization and transfer (16, 17, 43).

The presence of photosensitizers in fish oils depends on the method of processing and the source of fish used for rendering. Some species of fish, such as menhaden, which are primarily phytoplankton feeders, can easily contribute residual chlorophyll from their stomach contents when oils are rendered. Fish liver oils have a likelihood of containing residual heme compounds, and thus both oils would be generally more vulnerable to singlet oxygen autoxidation than oils recovered from dressed fish.

Although singlet oxygen hydroperoxidation is a non-specific initiation reaction, it provides an opportunity for siting hydroperoxides at otherwise unfavored positions on polyunsaturated fatty acids (44-46). In oxidizing cod liver and menhaden oils a compound that exhibits an odor reminiscent of vegetables and cooked celery which may contribute to green notes of oxidizing fish oils has been tentatively identified as 2-hydroxy-3-pentenalactone (32). The formation of

2-hydroxy-3-pentenalactone likely would require the siting of a hydroperoxide at the 5-position of DHA followed by reactions leading to the cyclization of the unsaturated 5-oxo-3-pentenoic acid. A siting of this hydroperoxide at the 5-position on DHA would not be favored in pentadienyl free radical autoxidation (44-46), but could be readily formed by a singlet oxygen mechanism.

Farmer-type autoxidation-derived fish flavors

The now classic Farmer-type hydrogen-abstraction initiation of free radical autoxidation accounts for a large portion of the nonenzymic oxidations of n-3 fatty acids (45). Because fish lipids contain substantial concentrations of EPA and DHA (47-48), they provide many allowed sites (18, 22, 45, 46, 49) of hydroperoxide formations, and thus can account for a large array of decomposition products. Oxidizing model systems of unsaturated methyl esters of fatty acids yielded monohydroperoxides, but also produce dihydroperoxides that are formed by cyclization of intermediate hydroperoxy radicals when suitable H-donating antioxidants are not present to quench the free radical reaction (45, 50, 51). Decomposition of monohydroperoxides of fatty acids in model systems yields a very different profile of lower molecular weight products than observed for similar decompositions of dihydroperoxides of the same fatty acids (45, 46).

Only a limited number of volatile products generated from n-3 fatty acids provide characterizing green, fishy or burnt flavors of oxidizing fish and fish oils. Allowed-site monohydroperoxide formations (18, 22, 45, 46, 49) and decompositions leading to some of the characterizing aldehydes via Farmer-type reactions in fish oils are summarized in Figure 1. As noted earlier, some aldehydes responsible for the green flavors and aromas in autoxidizing fish oils are common with those solely generated enzymically in freshly killed fish (23). Differences in concentrations of these and other compounds account for the overall aroma or flavor quality of fishery products exclusive of those contributed by microbial spoilage.

Of the aldehydes generated from the classic Farmer autoxidation mechanism, the 2,4,7-decatrienal isomers have been identified as being highly contributory to burnt/fishy or cod liver oil-like flavors (28, 32). Typically, both isomers exhibit the burnt/fishy flavor character, but the t,t,c-isomer can also contribute a green-fishy character to oxidizing fish oils at low concentrations (28, 32).

Directing effects of tocopherol-like antioxidants

Antioxidant systems for fish oils frequently contain alpha-tocopherol, but these systems and those based on phenolic antioxidants do not provide adequate protection against the formation of fishy flavors in fish oils. Swoboda and Peers (52-53) investigated a metallic/fishy flavor that developed only

in fractions of butteroil containing alpha-tocopherol and copper.
These workers reported that 1,5-octadien-3-one from the oxidation
of long-chain polyunsaturated n-3 fatty acids was responsible for
metallic notes, but the fishiness appeared to be contributed by
the 2,4,7-decatrienals or a combination of these compounds. This
relationship has been explored further in our laboratory through
mechanistic studies employing alpha-tocopherol and Trolox C which
is a synthetic tocopherol-type antioxidant (Figure 2).

Since alpha-tocopherol was an essential component of the
Swoboda and Peers system (52) which yielded metallic/fishy
flavors, its structural role in the directing of the flavor was of
interest. Either alpha-tocopherol (670 ppm) or Trolox C (1000
ppm) were added to steam deodorized menhaden oils (2 h at 130°C, 4
mm Hg; 32) and allowed to oxidize while exposed to air and
protected from the light at 65°C. As observed in previous studies
(54-56), the high level of alpha-tocopherol exhibited a prooxidant
effect whereas Trolox C possessed a distinct antioxidant effect in
oxidizing menhaden oils (57).

Both alpha-tocopherol and Trolox C are capable of donating a
hydrogen atom radical from the hydroxyl group of the chroman ring
(Figure 3) to quench free radicals in oxidizing lipid systems
(57-58). Differences in the stability of the chroman free radical
yielded by these two compounds account for the different
antioxidant properties exhibited. Although alpha-tocopherol
readily donates its hydrogen atom in the initial stage of free
radical quenching, the resulting chroman free radical is an
effective competitor for abstraction of hydrogen atoms from
unsaturated lipids or other hydroperoxides in oxidizing systems
(58). Consequently, this reverse process allows for the promotion
of oxidation by forming increased levels of free radicals. On the
other hand, Trolox C, having a strong electron withdrawing
carboxyl group substituted for the alkyl chain of alpha-tocopherol
(Figure 2), more readily undergoes the two step oxidation process
to form the corresponding oxidized lactone or quinone (Figure 3;
57). Formation of the quinone not only prevents the reversible
H-abstraction from occurring, but it also allows for the quenching
of an additional free radical in the process (Figure 3).

Headspace volatiles from the experimentally oxidized fish oils
were analyzed quantitatively by the method of Olfasdottir et al.
(59), and revealed that tocopherol-type antioxidants in samples
increased the formation of 1,5-octadien-3-ol and
1,5-octadien-3-one compared to a control (Figure 4). Trolox C had
a more pronounced effect than alpha-tocopherol. A proposed
mechanism for the enhanced formation of the two C_8 compounds is
illustrated in Figure 5, and involves an alignment of the pi-bond
clouds of the oxidized quinone with the n-3 end of the fatty acid
fragment. This alignment provides the desired geometry for the
transfer of the hydroxyl group to the 3-position of the 8-carbon
fragment. Because 1,5-octadien-3-ol is produced more abundantly
than the corresponding ketone, it appears that direct
hydroxylation is favored over an oxygen transfer to form
1,5-octadien-3-one. As indicated in Figure 5, a possibility

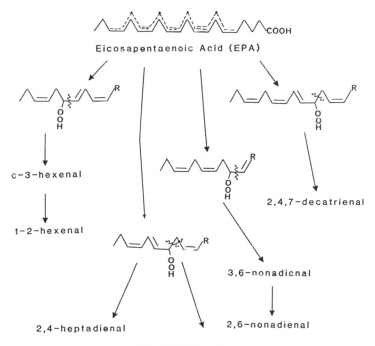

c-3-hexenal

t-2-hexenal

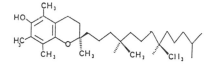

2,4,7-decatrienal

3,6-nonadienal

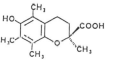

2,4-heptadienal

2,6-nonadienal

3,5-octadien-2-one

Figure 1.   Carbonyls associated with characterizing fish flavors produced from the autoxidation of eicosapentaenoic acid.

alpha-Tocopherol

Trolox C[R]

Figure 2.   Structures of tocopherol-type antioxidants evaluated in oxidizing fish oils.

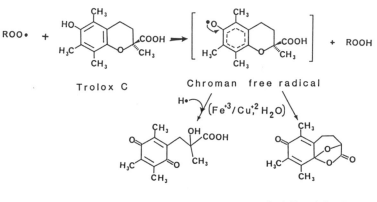

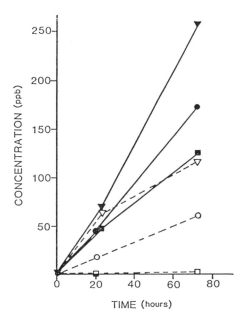

Figure 3. Antioxidant mechanism of Trolox C.

Figure 4. Formation of 1,5-octadien-3-ol and
1,5-octadien-3-one in the presence of tocopherol-type
antioxidants in menhaden oil under accelerated conditions.
[▼,▽ Trolox C; ●,○ alpha-tocopherol; ■,□ control; (— — —),
1,5-octadien-3-one; (————), 1,5-octadien-3-ol].

exists for the formation of the ketone from the alcohol through an oxidation reaction.

Theoretically, the alignment of the electron clouds of the molecules could facilitate both the hydroxyl group siting and the cleavage of the fatty acid backbone to provide the C_8 fragment, or alternatively the hydroxylation could occur after C_8 fragment has been released by an unrelated event. Since Trolox C is more readily oxidized to the quinone structure than alpha-tocopherol (57), the large concentration of unsaturated C_8 compounds produced in the presence of Trolox C supports a hypothesis that the directing action is more likely dependent on the formation of the oxidized chroman ring rather than the alignment character that may be provided by the alkyl chain.

When 20 ppm of Cu^{++} (as cupric palmitate) was added to deodorized cod liver oil, similar effects of alpha-tocopherol and Trolox C on the production of 1,5-octadien-3-ol and 1,5-octadien-3-one were observed except that even higher concentrations of the C_8 compounds resulted (Table I). This observation may be explained by the catalytic effect of metal ions in the formation of the oxidized quinone (57). Incorporation of Trolox C and copper caused the production of about four times the concentration of the unsaturated C_8 compounds compared to the sample containing alpha-tocopherol and copper. The sample containing Trolox C exhibited a strong metallic, vinyl ketone-like aroma quality (Table I) which probably was of the type described by Swoboda and Peers (52-53). Alpha-tocopherol and copper in fish oils, however, also allowed the development of high concentrations of the 2,4,7-decatrienals which resulted in an obscuring of the metallic note and the expression of strong cod liver oil-like aroma and flavor.

The ready donation of H-radicals from the hydroxyl group on the chroman ring quenches peroxy radicals formed on partially oxidized polyunsaturated fatty acids, and thus prevents intramolecular cyclizations that can lead to the formation of dihydroperoxides (22, 45, 50, 51). Thus, monohydroperoxide degradations lead to the fishy long-chain unsaturated aldehydes while diperoxide decompositions form shorter chain compounds (45-46) that contribute only to the general oxidized, painty flavor of oxidizing fish oils.

The H-donating character of the tocopherol-type compounds also causes a preferential formation of cis-trans rather than trans-trans monohydroperoxides that provide the direct precursors of the 2,4,7-decatrienals causing burnt/fishy flavors. A stepwise mechanism for the formation of t,t-hydroperoxide compared to t,c-hydroperoxide has been proposed by Porter et al. (51, 60). These researchers suggested that once the peroxy radical formed on the molecule, carbon-carbon bonds could rotate to establish the initial trans configuration (step 1; Figure 6). After bond rotation, it was proposed that the peroxy free radical could translocate along the carbon backbone (step 2, Figure 6) resulting in alternate siting at an allowed pentadienyl position on the

Table I. Odor and concentrations of selected compounds forming in oxidizing cod liver oils incorporating selected antioxidants and held at 65°C for 16.5 h

Sample description	1,5-C_8-3-ol	1,5-C_8-3-one	t,c,c-2,4,7-decatrienal	t,t,c-2,4,7-decatrienal	Odor description
			----------concentration (ppb)----------		
Control	65	tr[a]	320	400	green; fishy
670 ppm alpha-tocopherol + 20 ppm Cu^{++}	669	75	2710	1370	very cod liver oil-like; straw-like
1000 ppm Trolox C + 20 ppm Cu^{++}	3480	264	1200	390	metallic; vinyl ketone-like

[a]Trace.

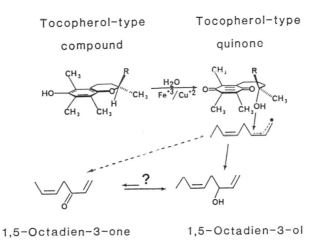

Tocopherol-type compound

Tocopherol-type quinone

1,5-Octadien-3-one   1,5-Octadien-3-ol

Figure 5. Proposed mechanism for the directed oxidation by tocopherol-type antioxidants for n-3 fatty acid fragments leading to the formation of C_8 compounds.

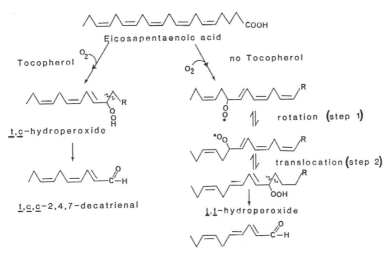

Eicosapentaenoic acid

Tocopherol

no Tocopherol

t,c-hydroperoxide

rotation (step 1)

translocation (step 2)

t,c,c-2,4,7-decatrienal

t,t-hydroperoxide

t,t,c-2,4,7-decatrienal

Figure 6. Formation of isomeric hydroperoxides of polyunsaturated fatty acids.

molecule. In the presence of strong H-donating antioxidants, the immediate quenching of peroxy radical would prevent the rotation of the carbon-carbon bond as well as the translocation of the peroxy radical on the unsaturated fatty acid, and thus minimize the formation of trans-trans hydroperoxide isomer formation (Figure 6; 41, 50, 60, 61).

The directed formation of the t,c,c-decatrienal isomer compared to the t,t,c-isomer in fish oil is illustrated by the ratio comparisons shown in Table II. In samples of cod liver oil containing either alpha-tocopherol plus copper or Trolox C plus

Table II. Ratios of selected C_8 and C_{10} carbonyls produced from oxidized cod liver oils containing tocopherol-like antioxidants

Sample description	Ratio of volatile compounds			
	C_8	t,c,c-C_{10}	C_8-ol	C_8-ol
	C_{10}	t,t,c-C_{10}	t,c,c-C_{10}	t,t,c-C_{10}
Control	0.09	0.80	0.20	0.16
670 ppm alpha-tocopherol + 20 ppm Cu^{++}	0.18	1.98	0.24	0.49
1000 ppm Trolox C + 20 ppm Cu^{++}	2.35	3.08	2.94	8.92

copper, at least twice the amount of the t,c,c-2,4,7-decatrienal isomer was produced as compared to the control. Since Trolox C exhibits high non-reversible H-donating antioxidant properties, its ability to more effectively direct toward the t,c,c-isomer is seen by an elevated isomer ratio of 3.08 in the presence of Trolox C (Table II). As noted earlier, low concentrations of the t,c,c-isomer of 2,4,7-decatrienal exhibit a more pronounced fishy/burnt flavor quality in oxidizing fish oils while the t,t,c-isomer causes more green-fishy, burnt character notes (28). Although substantial concentrations of the C_8 unsaturated compounds were produced in the fish oils containing Trolox C or alpha-tocopherol (Table I), the production of 2,4,7-decatrienals provided the underlying unpleasant, characterizing fish flavor notes.

Since both the C_8 and C_{10} unsaturated aldehydes are derived from the n-3 end of polyunsaturated fatty acids, the aroma and flavor profile of lesser oxidized fish oils would be expected to reflect the influences provided by the antioxidant system. Since Trolox C favors formation of C_8 unsaturated compounds compared to the unsaturated C_{10} compounds during the oxidation of fish oils (C_8:C_{10} ratio = 2.35; Table II), it is likely that less fishy aromas will result from its use, but metallic notes from the C$_8$ compounds may likewise cause problems with flavors. However, it may be possible to mask metallic flavors more readily that the fishiness caused by the 2,4,7-decatrienals.

Co-oxidation of carotenoids and fish lipids

Recent studies on salmon flavors revealed that a single compound appears to be responsible for the characterizing cooked salmon flavor (39). The cooked salmon flavor compound was found to have an extremely low threshold, and was initially detected only by odor assessment of a fraction eluting at I$_E$ of 9.6-9.7 on a Carbowax 20M packed column when headspace volatiles were analyzed from canned salmon meat. Accelerated oxidation of salmon oil did not yield salmon-like aromas before the development of fishy oxidized aromas. However, when salmon oil was coated onto Celite supports, and allowed to oxidize at room temperature, a distinct salmon-loaf-like aroma developed within 24 h after initiation of oxidation. A variety of supports were evaluated in model systems with salmon oil for their ability to produce the salmon aroma compound. Odor assessments of the oxidizing systems Table II indicated that a range of odors developed from salmon-loaf-like to oxidized fishy aromas, and only the Celite system provided the aroma.

The interaction of the carotenoid and the fatty acid fractions on Celite were both necessary for the odor development to occur. Studies designed to confirm an interaction of the carotenoid and fatty acid fractions in the development of salmon flavors showed that when carotenoid fractions from salmon oils were separated from the acylglycerol fraction by column chromatography, neither yielded a salmon-like aroma during oxidation (Table III). However, when the carotenoids and acylglycerols were recombined, the salmon aroma developed. Combinations of alternate sources of fish acylglycerols along with crayfish carotenoids revealed that the necessary component for salmon flavor development was the presence of carotenoids specifically derived from salmon oil (Table III). Such results strongly suggest that the compound is derived by co-oxidation of fish acylglycerols with salmon carotenoids, and that the precursor is located in the carotenoid fraction.

The GC-collected salmon flavor compound was analyzed by mass spectrometry and found to possess a molecular weight of 138 and mass fragments reminiscent of an alkyl furanoid-type structure. A compound fitting this type of molecular configuration was suggested as being ethyl-(3-methyl furyl)-ketone which would accomodate the analytical evidence (39). Formation of such a compound from salmon carotenoids would likely require an allene grouping in a carotenoid which could lead to the formation of a ring compound.

In addition to the directed oxidation process responsible for the characteristic salmon flavor compound, carotenoids also were observed to influence oxidation of polyunsaturated fatty acids in salmon oil systems (39). Typically, lipid oxidation of oils high in n-3 fatty acids produce isomeric 2,4-heptadienals as the major

Table III.   Carotenoid and fatty acid
contributions to cooked salmon flavor[a]

Oil/Celite System	Overall Aroma Quality
From Salmon Oil	
Carotenoid	unsaturated hydrocarbon
Acylglycerol	rancid fish
Carotenoid + Acylglycerol	cooked salmon loaf
From Other Sources	
Salmon carotenoid + Menhaden oil	cooked salmon loaf
Crayfish carotenoid + Salmon acylglycerol	no salmon aroma
Salmon acylglycerol + Menhaden oil	rancid fish

[a]From Josephson, 1987 (39).

autoxidation products, and the isomeric 3,5-octadien-2-ones are
present as less abundant breakdown products (30, 31, 61). Studies
employing salmon oil containing carotenoids as well as
fractionated salmon oil devoid of carotenoids showed that the
ratio of the 2,4-heptadienal isomers and the 3,5-octadien-2-one
isomers were very different in the two systems. Although all of
the isomers of both of these lipid oxidation products are derived
from the n-7 hydroperoxide of n-3 fatty acids, a very distinct
difference in the breakdown of this hydroperoxide occurred in the
presence and absence of the carotenoid fraction. These particular
oxidation products are not considered to provide characterizing
fish flavors, but their varying contributions would likely be
notable in the overall quality of fish aromas and flavors.

Role of oxidatively-produced saturated fatty acids in fish flavors

We originally believed that the short chain saturated fatty acids
in oxidizing fish lipids contributed to burnt/fishy flavors.
Saturated fatty acid concentrations ($C_4 - C_9$) measured by
volatile headspace analysis (32, 59) reached levels as high as
3 ppm in highly oxidized fish oils (32). Flavor threshholds for
these short n-chain fatty acids in oil systems in the literature
(>.66 ppm, 63) indicate that they could contribute notes to
oxidizing fish oils. However, studies designed to document the
role of short chain acids as flavor compounds detracting from the
flavor quality of fish oils did not confirm earlier beliefs.
    Short chain fatty acids are formed by peracid oxidations of
autoxidatively-derived n-alkanals (63), and the incorporation of
the peracid inhibitor, dilauryl thiodipropionate (63), into
deodorized fish oils was investigated as a means of preventing the
formation of fishy/burnt flavors. Although results of these
studies showed substantial reduction of these acids to
concentrations well below threshold (<10 ppb), burnt/fishy flavors

were still pronounced and unaffected. When concentrations up to 20 ppm of either butyric, pentanoic, hexanoic, heptanoic and octanoic acids were added to deodorized fish oils and bland vegetable oils, only butyric acid gave a weak cheese-like, buttery aroma (32). Therefore, it was concluded that short n-chain fatty acids found in oxidizing fish oils were of insignificant concentrations to contribute characterizing burnt/fishy flavors and aromas.

Summary

Although it is well established that flavors generated from oxidative processes in fish and fishery products are difficult to control, some mechanisms for altering or directing predominant oxidative pathways to alternate flavors and aromas exist. We have only begun to unravel these complex relationships in fish flavor systems, but the results provide encouragement toward the goal of one day controlling the distinctly fishy flavors produced from the oxidation of long-chain polyunsaturated fatty acids.

Acknowledgments

This research was supported by the College of Agricultural and Life Sciences and the Sea Grant College Program, Federal Grant No. NA84A-D-0065, Project AS/A-8, University of Wisconsin-Madison.

Literature Cited

1. Stansby, M.E. 1976. Fish Oils: Their Chemistry, Technology, Stability, Nutritional Properties and Uses; Stansby, M.E. Ed. AVI Publish. Co., Westport, CT 1967; pp. 171-192.
2. Stansby, M.E. J. Am. Oil Chem. Soc. 1971, 40, 820.
3. Stansby, M.E. J. Am. Oil Chem. Soc. 1978, 55, 238.
4. Josephson, D.B., Lindsay, R.C. and Olafsdottir, G. Seafood Quality Determination; Kramer, D.E. and Liston, J. Eds. Elsevier Science Publ., Amsterdam, 1986; pp. 27-47.
5. Lindsay, R.C., Josephson, D.B. and Olafsdottir, G. In Seafood Quality Determination; Kramer, D.E. and Liston, J. Eds. Elsevier Science Publ., Amsterdam, 1986; pp. 221-234.
6. Kubota, K. and Kobayashi, A. J. Agric. Food Chem. 1988, 36, 121.
7. Gunstone, F.D. J. Am. Oil Chem. Soc. 1984, 61, 441.
8. Tappel, A.L., Boyer, P.D. and Lundberg, W.O. J. Biol. Chem. 1952, 199, 267.
9. Ingraham, L.L. Comprehensive Biochemistry; Florkin, M. and Stotz, E.H. Eds. Elsevier Publ. Co., New York, 1966; pp. 424-446.
10. German, J.B., Bruckner, G.G. and Kinsella, J.E. Biochem. Biophys. Acta. 1986, 875, 12.
11. German, J.B. and Kinsella, J.E. J. Agric. Food Chem. 1985, 33, 680.
12. Foote, C.S. Free Radicals in Biology; Vol. II. Pryor, W.A. Ed. Academic Press, New York, 1976, pp. 85-133.

13. Adams, W.   Chem. Ztg. 1975, 99, 142.
14. Gollnick, K.   1968. Advanced Photochemistry; Noyes, W.A.,
     Hammond, G.S. and Pitts, N.J. Eds. Interscience Publ., New
     York, 1968, 6, 1.
15. Livingston, R. Autoxidation and Antioxidants; Vol. I.
     Lundberg, W.O., Ed. Wiley and Sons Publ., N.Y., 1961, pp.
     249.
16. Nawar, W.W. Food Chemistry; Fennema; O.R. Ed. Marcel
     Dekker, Inc., New York, 1985, pp. 139-245.
17. Korycka-Dahl, M.B. and Richardson, T. CRC Crit. Rev. in
     Food Sci. and Nutri. Vol. 10, 1978, pp. 209-238.
18. Frankel, E.N. Prog. Lipid Res., 1980, 19, 1.
19. Grosch, W. Food Flavors, Part A: Introduction; Morton,
     I.D. and Macleod, A.J. Eds. Elsevier Scientific Publish.
     Co., New York, 1982, pp. 325-398.
20. Lillard, D.A. Lipids as a Source of Flavor; Supran, M.K.
     Ed. ACS Symposium Series #75. Washington, D.C., 1978, pp.
     68-80.
21. Esterbauer, H. Free Radicals, Lipid Peroxidation and
     Cancer; McBrien, D.C.H. and Slater, T.F. Eds. Academic
     Press, London, 1982, pp. 101-122.
22. Chan, H.W.-S., and Coxon, D.T. Autoxidation of Unsaturated
     Lipids; Chan, H.W.-S. Ed. Academic Press, New York, 1987,
     pp. 17-50.
23. Josephson, D.B. and Lindsay, R.C. Symposium on
     Biogeneration of Aromas; Am. Chem. Soc., Chicago, IL, 1986,
     pp. 201-219.
24. Josephson, D.B., Lindsay, R.C., and Stuiber, D.A. J. Food
     Sci. 1985, 50, 5.
25. Josephson, D.B., Lindsay, R.C., and Stuiber, D.A. J. Agric.
     Food Chem. 1984, 32, 1344.
26. Josephson, D.B., Lindsay, R.C., and Stuiber, D.A. J. Agric.
     Food Chem. 1984, 32, 1347.
27. Josephson, D.B., Lindsay, R.C., and Stuiber, D.A. J. Agric.
     Food Chem. 1983, 31, 326.
28. Meijboom, P.W. and Stroink, T.B.A. J. Am. Oil Chem. Soc.
     1972, 49, 555.
29. Badings, H.T. J. Am. Oil Chem. Soc. 1973, 50, 334.
30. McGill, A.S., Hardy, R., Gunstone, F.D. J. Sci. Food Agric.
     1977, 28, 200.
31. Josephson, D.B., Lindsay, R.C. and Stuiber, D.A. Can. Inst.
     Food Sci. Technol. J. 1984, 17, 187.
32. Karahadian, C. Ph.D. Thesis, University of
     Wisconsin-Madison, 1988.
33. Slabyj, B.M. and Hultin, H.O. J. Food Sci. 1984, 49, 1392.
34. Slabyj, B.M. and Hultin, H.O. J. Food Sci. 1982, 47, 1395.
35. Shewfelt, R.L., McDonald, R.E. and Hultin, H.O. J. Food
     Sci. 1981, 46, 1297.
36. McDonald, R.E., Kelleher, S.D. and Hultin, H.O. J. Food
     Biochem. 1979, 3, 125.
37. Kanner, J., Harel, S. and Hazan, B. J. Agric. Food Chem.
     1986, 34, 506.
38. Kanner, J. and Kinsella, J.E. J. Agric. Food Chem. 1983,
     31, 370.

39. Josephson, D.B. Ph.D. Thesis, University of Wisconsin-Madison, 1987.

40. Grosch, W. and Laskawy, G. J. Agric. Food Chem. 1975, 23, 791.

41. Wardale, D.A. and Gallaird, T. Phytochem. 1975, 14, 2323.

42. Gardner, H.W. Advances in Cereal Science and Technology; Vol. IX; Pomeranz, Y. Ed. Am. Assoc. of Cereal Chemists, St. Paul, MN 1988, pp. 161-215.

43. Clements, R.H., van der Engh, R.H., Frost, D.J., Hoogenhout, K., Nooi, R.F. J. Am. Oil Chem. Soc. 1973, 50, 325.

44. Chan, H.W.-S. J. Am. Oil Chem. Soc. 1977, 54, 100.

45. Frankel, E.N. J. Am. Oil Chem. Soc. 1984, 61, 1908.

46. Grosch, W. Autoxidation of Unsaturated Lipids; Chan, W.W.-S. Ed. Academic Press, New York, 1987, pp. 95-139.

47. Ackman, R.G. Objective Methods for Food Evaluation, Proc. of Symposium; Nat. Acad. of Sci., Washington, D.C. 1976, pp. 103-118.

48. Kinsella, J.E. Seafoods and Fish Oils in Human Health and Disease; Marcel Dekker, Inc. New York, 1987.

49. Chan, H.W.-S., Coxon, D.T., Peers, K.E. and Price, K.R. Food Chem. 1982, 9, 21.

50. Peers, K.P., Coxon, D.T. and Chan, H.W.-S. 1981. J. Sci. Food Agric. 1981, 32, 898.

51. Porter, N.E., Lehman, L.S., Wever, B.A., Smith, K.J. J. Am. Chem. Soc. 1981, 103, 6447.

52. Swoboda, P.A.T. and Peers, K.E. J. Sci. Food Agric. 1977, 28, 1010.

53. Swoboda, P.A.T. and Peers, K.E. J. Sci. Food Agric. 1977, 28, 1019.

54. Koskas, J.P., Cillard, J. and Cillard, P. J. Am. Oil Chem. Soc. 1984, 61, 1466.

55. Cillard, J., Cillard, P. and Cormier, M. J. Am. Oil Chem. Soc. 1980, 57, 255.

56. Cillard, J., Cillard, P., Cormier, M. and Girre, L. J. Am. Oil Chem. Soc. 1980, 57, 252.

57. Cort, W.M., Scott, J.W., Araujo, M., Mergens, W.J., Cannalonga, M.A., Osadca, M., Harley, J.H., Parrish, D.R. and Pool, W.R. J. Am. Oil Chem. Soc. 1975, 52, 174.

58. Terao, J. and Matsushita, S. Lipids 1986, 21, 25.

59. Olafsdottir, G., Steinke, J.A. and Lindsay, R.C. J. Food Sci. 1985, 50, 1431.

60. Porter, N.A., Wever, B.A., Weener, H. and Khan, J.A. J. Am. Chem. Soc. 1980, 102, 5597.

61. Yu, T.C., Day, E.A. and Sinnhuber, R.O. J. Food Sci. 1961, 26, 192.

62. Siek, T.J., Albin, I.A., Sather, L.A. and Lindsay, R.C. J. Food Sci. 1969, 34, 264.

63. Karahadian, C. and Lindsay, R.C. J. Am. Oil Chem. Soc. 1988, in press.

RECEIVED October 31, 1988

Chapter 7

Kinetics of Formation of Alkylpyrazines

Effect of Type of Amino Acid and Type of Sugar

M. M. Leahy[1] and G. A. Reineccius[2]

[1]Ocean Spray Cranberries, Inc., R & D Building, Bridge Street,
Middleboro, MA 02346
[2]Department of Food Science, University of Minnesota, St. Paul, MN 55108

Pyrazines are heterocylic, nitrogen-containing
compounds important to the flavor of many foods.
Prior studies relating to the effects of type of amino
acid and type of sugar on the formation of pyrazines
have yielded contradictory results. This study
investigates the effects of type of amino acid and
type of sugar on the kinetics and distribution pattern
of pyrazines formed. The amino acids, lysine and
asparagine, and the sugars, glucose, fructose and
ribose were chosen for this study. One-tenth molar
sugar/amino acid solutions buffered at pH 9.0 were
heat-processed. Samples were analyzed using a
headspace concentration capillary gas chromatographic
technique with nitrogen-selective detection. Rate of
pyrazine formation fit pseudo zero order reaction
kinetics. Effects of amino acid and sugar types on
activation energies, yields and relative distributions
of pyrazines are discussed.

Pyrazines are heterocyclic, nitrogen-containing compounds
important to the flavor of many foods. Alkylpyrazines have often
been found in heated foods and have been characterized as having
roasted, toasted, nutty flavor notes. Some excellent reviews have
previously detailed the presence of pyrazines in a great variety
of foods. Maga and Sizer (1, 2) published the first comprehensive
reviews on pyrazines in foods. They reviewed the occurrences of
numerous pyrazines in a wide variety of foods, pyrazine isolation,
concentration, separation and identification techniques, pyrazine
flavor properties, pyrazine mechanisms of formation and pyrazine
synthesis techniques. Since then several others (3 - 10) have
reviewed progress in pyrazine research. Understanding both the
effects of various parameters on the kinetics of the formation of
pyrazines and the mechanism of formation may allow for the

0097–6156/89/0388–0076$06.00/0

optimization of pyrazine production in both foods and in reaction flavors.

Alkylpyrazines are most commonly found in roasted and toasted foods and are believed to form as a result of the Maillard browning reaction (11). Several researchers have proposed mechanisms for alkylpyrazine formation in various carbohydrate/amine systems (12 - 21). These pathways generally involve the formation of aminocarbonyl fragments which condense, yielding dihydropyrazines or hydroxy dihydropyrazines. These in turn yield pyrazines through oxidation (17) or dehydration reactions (18, 19). Aminocarbonyl fragments result through various Maillard reaction pathways. Some researchers postulate mechanisms in which free ammonia formed as a result of amino acid decomposition reacts with sugars and sugar fragments yielding alkylpyrazines (15, 16). Others have proposed mechanisms by which sugars and amino acids condense through the generalized Hodge Maillard reaction scheme (11, 12, 13) and Strecker degradation of amino acids with dicarbonyl fragments (17). Shibamoto and Bernhard (18) proposed the most detailed scheme of pyrazine formation pathways in sugar-ammonia model systems, involving α-aminocarbonyl intermediates which condense to form alkylpyrazines.

Various researchers have investigated factors affecting both yields of individual pyrazines and their distributions, including source of nitrogen and source of carbon. Often these studies have yielded seemingly contradictory results. The effect of source of nitrogen on pyrazine formation was initially investigated by Newell et al. (13). They reacted various amino acids with glucose and demonstrated that qualitatively the same volatile pyrazine compounds were produced regardless of the amino acid employed as the nitrogen source. Van Praag et al. (15) obtained similar results in reacting glycine, leucine, isoleucine, valine and alanine with D-fructose. Koehler et al. (16) observed formation of a similar series of pyrazines when reacting glucose with asparagine, glutamine, glutamic acid and aspartic acid. Mostly alkylated pyrazines were found, with only traces of unsubstituted pyrazine. However, in reacting ammonium chloride with glucose, unsubstituted pyrazine was the main pyrazine observed with only traces of alkylated pyrazines being found. Also, different pyrazine product distributions and total pyrazine concentrations were obtained for the different amino acids. Wang and Odell (14) reacted gycerol with alanine and glycine. Similar series of pyrazines resulted, however total yields and distributions were different. Most recently, Wong and Bernhard (20) demonstrated that the distribution and yield of pyrazines formed in the reactions of glucose with ammonium hydroxide, ammonium formate, ammonium acetate, glycine and monosodium glutamate depended strongly on the nature of nitrogen source.

The effect of source of carbohydrate on pyrazine formation was first investigated by van Praag et al. (15). The same series of pyrazine compounds was isolated from a reaction mixture of fructose or glucose with ammonia. Koehler et al. (16) found that the carbohydrate source affected the total yield of pyrazines formed. In reacting 10 mmoles each of a sugar with ammonium chloride, total yield of pyrazine was 59 μmoles with glucose and

195 μmoles with fructose. Koehler and Odell (22) monitored
production of methylpyrazine and dimethylpyrazine in reacting
asparagine with glucose, fructose, sucrose and arabinose. They
found the carbon source to affect both total yield and relative
distribution. In a more comprehensive study, Shibamoto and
Bernhard (18) reacted ammonia with several sugars. Reaction
systems consisted of a solution of 8M ammonium hydroxide and 1M
carbohydrate, heated at 100°C for 2 h. They concluded that
pentoses give greater total yields of pyrazines than hexoses, and
yields from mannose, glucose and fructose are about equal with
galactose giving slightly lower yields. Distribution patterns
were essentially similar for the pentoses and hexoses, except
aldoses gave more unsubstituted pyrazine than the ketose,
fructose. They concluded that epimers, diastereomers and
enantiomers gave identical pyrazine pattern distributions,
contradicting the work of Koehler and Odell (22).

Our study was conducted to determine the effects of type of
amino acid and type of sugar on both the kinetics and the
distribution pattern of alkylpyrazines formed. For investigating
the effect of type of amino acid, this work focused on two amino
acids, lysine and asparagine. Lysine was chosen because it
contains two amino groups available for reaction. Asparagine was
chosen because Koehler and Odell (22) reported that yields of
pyrazines were greatest when this amino acid was reacted with
glucose, verses alanine, glycine or lysine. For investigating the
effect of sugar, glucose, fructose and ribose were chosen for this
study. Glucose and fructose were chosen to provide two common
hexoses in foods, one an aldose and one a ketose. The pentose
ribose was chosen because of its reportedly high degree of
reactivity in meat reaction flavor systems (Dwidedi, 23).

The present investigation used a headspace concentration
capillary gas chromatographic technique with nitrogen-phosphorus
detection. Advantages of this technique are that the procedure
was rapid (about 30 minutes), potential for artifact formation is
minimized and sample requirements are small (15 ml).

Experimental Section

The amino acids, sugars, borate salts and solvents were all
reagent grade, obtained from commercial sources. The amino
acid-sugar combinations investigated in the present study included
asparagine-fructose, asparagine-glucose, lysine-fructose,
lysine-glucose and lysine-ribose at concentrations of 0.1M for
both amino acid and sugar, in a pH 9.0 0.1M borate buffer (24).
Ten ml of each solution were heated in Teflon-capped 25 mm (o.d.)
x 150 mm Pyrex test tubes in a water bath at 75, 85, and 95°C for
up to 24 h. Samples were taken at 7 to 8 time intervals.
Eighteen to 20 total samples per temperature were analyzed.

Although the amino acid/sugar solutions were buffered in an
effort to maintain pH, a drop in pH was encountered with
increasing reaction times. Therefore, after heat treatment, each
sample was adjusted to pH 9.0 with 0.1N NaOH. One ml of a
solution containing 2-methoxypyrazine in distilled water (2 ppm)
was added as an internal standard. Final sample volume was 15 ml.
Pyrazines were then isolated, separated and quantified using an

automated headspace concentration sampler (Hewlett Packard 7675A
Purge and Trap) coupled to a Hewlett Packard 5880A gas
chromatograph with nitrogen-phosphorus detection. The 15 ml
sample was attached to the purge and trap sampler and purged for
10 min with hydrogen at a flow rate of 90 ml/min. Volatiles were
adsorbed on a 4" by 1/4" o.d. stainless steel precolumn packed
with Tenax (Hewlett Packard Co., Avondale, PA). At the end of the
purge period, the sample was removed from the sampler before
manually switching to the desorb cycle. Elution of the
concentrated organic volatiles from the Tenax precolumn onto the
GC column was accomplished by heating the precolumn to 180°C with
a hydrogen flow of 90 ml/min. The combined flow of hydrogen and
volatiles was split 50:1 and passed onto the GC column, a 25 m by
0.32 mm (i.d.) DB 225 fused silica capillary column (J & W
Scientific Inc., Rancho Cordova, CA). The column head pressure
was maintained at 15 psig which provided a linear velocity of 45
cm/sec and a flow rate of 4 ml/min. The volatiles were desorbed
for 3 min from the Tenax column onto the chromatographic column.
During this period, the volatiles were cold-trapped at the head of
the chromatographic column by immersing a 20 cm loop of the column
in a 2 3/8" i.d. x 4 1/2" Dewar flask containing liquid nitrogen.
The oven temperature was isothermal at 50°C for the run, with a
post value of 200°C for 1.5 min. The injection port and NPD
temperatures were 225°C and 280°C, respectively. Figure 1 gives a
typical chromatogram obtained by this procedure.

Quantification of the pyrazines was accomplished using an
internal standard method. 2-methoxypyrazine was chosen as
internal standard due to its similar physical properties
(MW, solubility) to the pyrazines of interest and because
methoxypyrazines have never been reported to form as a result of
the Maillard reaction. The amounts of each pyrazine present in
the sample were determined by the relationship:

$$\text{Amt. C} = \frac{\text{Amt. ISTD}}{\text{RF}} \times \frac{\text{AC C}}{\text{AC ISTD}} \qquad (1)$$

where RF = response factor of the compound, C,
           relative to the internal standard,
           ISTD
      AC = area count
      Amt. = amount in µg/ml

Response factors were empirically determined by adding known
amounts of each compound to 15 ml of pH 9.0 borate buffer,
followed by a purge of the sample under usual conditions of
analysis.

Pyrazine peak identification was initially accomplished by
cochromatography with standards, (Pyrazine Specialties, Atlanta,
GA) then further confirmed by gas chromatography/mass spectrometry
using the following sample, equipment and operating conditions. A
10 ml sample of 0.1M glucose-lysine heated at 95°C for 6 h was
analyzed. Following pH adjustment to 9.0 with 0.1N NaOH, the
volume was brought up to 15 ml. The sample was analyzed using the
Hewlett Packard 7675A Purge & Trap sampler interfaced with a Carlo

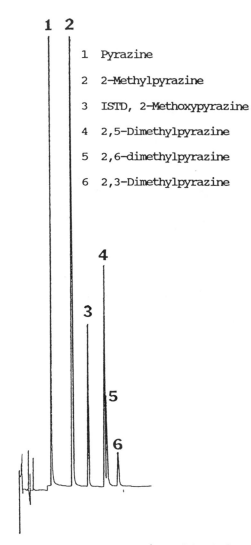

1   Pyrazine

2   2-Methylpyrazine

3   ISTD, 2-Methoxypyrazine

4   2,5-Dimethylpyrazine

5   2,6-dimethylpyrazine

6   2,3-Dimethylpyrazine

Figure 1.   Chromatogram of pyrazines detected under standard
conditions of analysis

Erba gas chromatograph connected to a Kratos MS 25 mass spectrometer. Helium was used as carrier at a pressure of 90 kPa. The same column was used for routine pyrazine analysis, a 25 m x 0.32 mm i.d. DB 225 fused silica capillary column. The run was isothermal at 50°C with a post value of 200°C for 2 min. Spectra were recorded at 70 electron volts. The spectra were compared to published spectra (25) as well as those obtained from reference standards.

The kinetics of the formation of pyrazines were determined using the basic equation for the rate of change of A with time:

$$\frac{dA}{d\theta} = kA_n \qquad (2)$$

where A = concentration of pyrazine (ppm)
θ = time (h)
k = rate constant
n = reaction order

Integrating this equation between A_0, the concentration of A at time zero, and A, the concentration of A at time 0, yields

$$A = A_0 + k\theta \qquad (3)$$

for a zero order reaction. This implies that the rate of formation of A is constant with time and independent of the concentration of reactants. For a first order reaction this yields the relationship:

$$\ln A = \ln A_0 + k\theta \qquad (4)$$

In this case, the rate for formation of A is dependent on the concentration of reactants remaining. Reactions in foods have been found to follow pseudo zero or first order kinetics (26). One is generally safer when discussing reaction orders in foods in using the term "pseudo", due to the complexity of the system. Pseudo reaction orders in foods are generally assigned because a high correlation (r^2) for a mathematical relationship between formation of product and time exists.

The formation of pyrazines appears to better fit a pseudo zero order reaction rather than first order reaction. Plotting concentrations of pyrazines formed versus time of reaction gave the better fit of the line, usually with a coefficient of determination (r^2) of greater than 0.95. For a pseudo first order reaction, a curve rather than a line was obtained. General least squares analysis of the data was used to compute rate constants (27). Two zero points were used for each regression. Duplicate samples were tested at the early sampling times vs. triplicate samples at later times, as variations in concentration among replicates increased with increased reaction time. Each data point collected was treated separately in the regression analyses.

Activation energies for the formation of pyrazines were calculated using the Arrhenius relationship, which relates the rate constant, k, to temperature:

$$k = k_O - E_a/RT \qquad\qquad (5)$$

where k_O = a pre-exponential (absolute) rate constant
E_a = activation energy in kcal/mole
R = gas constant, 1.986 cal/mole $^O K$
T = temperature in $^O K$

Results and Discussion

Table I lists the regression data for the formation of pyrazines
with time for the different sugar-amino acid combinations. An
increase in rates of formation occurred with an increase in
temperature. The effect of temperature on the formation of
pyrazine in the lysine-glucose system is graphically depicted in
Figure 2. Formation of pyrazines best fit a pseudo zero order
reaction, with coefficients of determination usually greater than
0.90. This suggests that the rate of formation of pyrazines is
independent of reactant concentration. Using 0.1M solutions of
both reactants (sugar and amino acid), concentrations of reactants
are quite high, especially relative to pyrazines formed and
reactants consumed. Although the rate of formation must be a
function of reactant concentration, it may not be apparent due to
the relatively high reactant/product ratio, multiplicity of steps
in pyrazine formation and competing side reactions. Pigment
formation in the Maillard reaction, which is also a multi-step
process, has also been shown to exhibit pseudo zero order kinetics
by many researchers, including Labuza et al. (28) and Warmbier et
al. (29) when reactant concentrations were not limiting for the
rate of formation of brown pigment. Loss of reactants has
generally been shown to exhibit first order kinetics (29) as well
as formation of Amadori compounds, which is essentially a single
step process as far as reactants are concerned (30).

Activation energies for alkylpyrazine formation were
calculated from the slope of Arrhenius plots, ranging from 27 to
45 kcal/mole (see Table II). Activation energies have been
reported for other aspects of the Maillard reaction. The
activation energy for browning as measured for pigment production
ranges from 15.5 kcal/mole for a glycine-glucose system (31) to 33
kcal/mole for a solid intermediate moisture model food system
(29). In the same system, Warmbier et al. (29) reported the
activation energies for both lysine and glucose loss to be 25
kcal/mole. The current research indicates that the activation
energies for pyrazine formation are higher, suggesting a different
rate-controlling step.

The mean activation energies for the dimethylpyrazines are
slightly higher than those of the unsubstituted and
2-methylpyrazines. An analysis of variance was performed treating
the five sugar-amino acid substrate combinations as a block and
pyrazine, 2-methylpyrazine and 2,5-dimethylpyrazine as treatments.
No difference in response was found among the substrates at a 95%
level of significance. A difference was found to exist among
treatments, at an α of 0.05. Duncan's multiple range test was
performed to determine which activation energies differed at the
same level of significance (0.05). It was found that the

Table I. Regressions for the effect of type of
sugar and amino acid on the formation of pyrazines

Model System	Temperature	k (ppm/hr)	k_0 (intercept)	Number of samples	r^2
LYSINE-GLUCOSE					
pyrazine	95 C	3.596	0.0596	22	0.994
	85 C	0.490	0.458	22	0.960
	75 C	0.214	0.279	20	0.965
2-methyl-pyrazine	95 C	2.837	-0.104	22	0.995
	85 C	0.422	0.142	22	0.967
	75 C	0.159	0.0910	20	0.941
2,5-di-methyl-pyrazine	95 C	0.186	-0.0457	20	0.995
	85 C	0.0247	-0.00604	22	0.985
	75 C	0.00668	-0.00536	16	0.942
2,3-di-methyl-pyrazine	95 C	0.0229	-0.00569	16	0.948
	85 C	0.00309	-0.00173	18	0.978
	75 C	0.000677	-0.00860	12	0.958
LYSINE-FRUCTOSE					
pyrazine	95 C	1.359	-0.226	20	0.995
	85 C	0.395	-0.0991	22	0.995
	75 C	0.134	-0.103	21	0.963
2-methyl-pyrazine	95 C	1.105	0.301	20	0.995
	85 C	0.388	0.119	22	0.945
	75 C	0.116	-0.0343	21	0.968
2,5-di-methyl-pyrazine	95 C	0.175	0.0488	20	0.945
	85 C	0.0376	0.0211	20	0.951
	75 C	0.00779	0.00570	15	0.935
2,3-di-methyl-pyrazine	95 C	0.0441	0.0127	16	0.954
	85 C	0.117	0.00328	16	0.969
LYSINE-RIBOSE					
pyrazine	95 C	3.488	0.530	21	0.974
	85 C	0.992	0.764	22	0.916
	75 C	0.310	0.747	22	0.882
2-methyl-pyrazine	95 C	5.364	1.380	19	0.957
	85 C	1.768	0.914	22	0.929
	75 C	0.440	0.350	22	0.952

Continued on next page.

Table 1 (cont'd)

Model System	Temperature	k (ppm/hr)	k_o (intercept)	Number of samples	r^2
LYSINE-RIBOSE					
2,5-di-	95 C	0.152	0.0265	18	0.975
methyl-	85 C	0.0440	0.0204	17	0.933
pyrazine	75 C	0.0128	0.00717	18	0.984
2,3-di-	95 C	0.0166	0.00127	17	0.973
methyl-	85 C	0.00427	0.000369	16	0.958
pyrazine					
ASPARAGINE-GLUCOSE					
pyrazine	95 C	0.103	-0.0371	22	0.987
	85 C	0.0422	-0.0385	22	0.941
	75 C	0.00981	-0.0187	22	0.947
2-methyl-	95 C	0.442	-0.215	22	0.992
pyrazine	85 C	0.179	-0.242	22	0.926
	75 C	0.0343	-0.102	22	0.942
2,5-di-	95 C	0.0871	-0.0117	22	0.982
methyl-	85 C	0.0202	-0.00255	22	0.968
pyrazine	75 C	0.00455	-0.0153	14	0.863
2,6-di-	95 C	0.0997	-0.0875	21	0.979
methyl-	85 C	0.0188	-0.0294	11	0.821
pyrazine					
2,3-di-	95 C	0.00231	-0.000192	20	0.992
methyl-					
pyrazine					
ASPARAGINE-FRUCTOSE					
pyrazine	95 C	0.0227	-0.0135	20	0.904
	85 C	0.00653	-0.00631	20	0.881
	75 C	0.00266	-0.00328	20	0.947
2-methyl-	95 C	0.632	-0.331	22	0.927
pyrazine	85 C	0.163	-0.197	22	0.926
	75 C	0.0374	-0.0928	22	0.896
2,5-di-	95 C	0.487	-0.107	22	0.983
methyl-	85 C	0.122	-0.0192	22	0.995
pyrazine	75 C	0.0261	-0.0323	20	0.940
2,6-di-	95 C	0.541	-0.248	20	0.960
methyl-	85 C	0.152	-0.104	20	0.980
pyrazine	75 C	0.0356	-0.0547	18	0.921

activation energies for 2,5-dimethylpyrazine were significantly higher than those of pyrazine and 2-methylpyrazine. However, in performing an analysis of variance the assumption must be made of constant variance for all analyses. This assumption was found to be violated in determining the variances on the regressions from which the activation energies were calculated. One might anticipate these results, since three points were used in the Arrhenius plots. Therefore, further research in this area is necessary to determine the significance of these findings.

Table II. Activation energies for formation
of pyrazines in 0.1M sugar-amino acid systems

PYRAZINE	E_a in kcal/mole
lys-glu	35.8
lys-fruc	29.5
lys-rib	30.8
asp-glu	30.0
asp-fruc	27.3
2-METHYLPYRAZINE	
lys-glu	36.6
lys-fruc	28.7
lys-rib	31.9
asp-glu	32.6
asp-fruc	36.0
2,5-DIMETHYLPYRAZINE	
lys-glu	42.3
lys-fruc	39.6
lys-rib	31.5
asp-glu	37.6
asp-fruc	37.3
2,6-DIMETHYLPYRAZINE	
asp-fruc	34.7
2,3-DIMETHYLPYRAZINE	
lys-glu	44.8

The effect of type of amino acid and sugar on the formation of pyrazines was investigated by comparing total yields and relative product distributions of sugar-amino acid systems heated at 95°C for 2 h, as seen in Table III and Figure 3. The type of amino acid reacted had a pronounced effect on total yield of pyrazines produced. A greater yield always resulted with lysine than with asparagine, especially when reacted with glucose. The lysine-glucose system yielded 13.1 ppm total pyrazines while the asparagine-glucose resulted in 0.74 ppm. A system consisting of 0.1M cysteine and glucose, pH 9.0, was also reacted under the same conditions. After 2 h of reacting at 95°C, no pyrazines were detected.

The effect of sugar on total yields of pyrazines was also investigated. In the lysine-sugar systems, total yield was greatest with the pentose, ribose, with 20 ppm total pyrazines produced. Glucose resulted in 13 ppm, versus fructose with 5.7 ppm total pyrazines. With asparagine, the effect of hexoses on

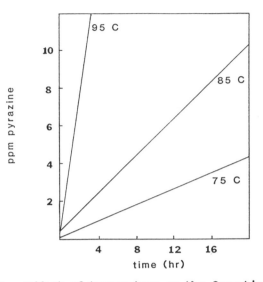

Figure 2. Effect of temperature on the formation of pyrazine

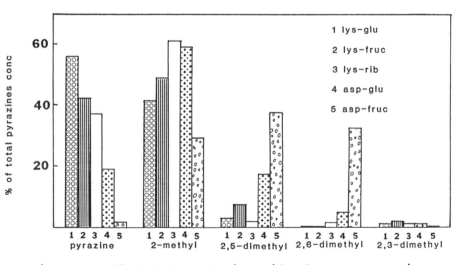

Figure 3. Effect of type of amino acid and sugar on pyrazine
distribution - 2 hr at 95°C

total yield showed reverse results, with fructose having a greater total yield than glucose (2.2 ppm versus 0.74 ppm).

Table III. Effect of type of amino acid and type of sugar on pyrazine distributions, 2 h treatment at 95°C

	L-G*	L-F*	L-R*	A-G*	A-F*
			%		
pyrazine	55.8	42.3	36.8	18.7	1.0
2-methylpyrazine	41.5	48.6	61.1	58.8	29.1
2,5-dimethylpyrazine	2.4	7.3	1.6	17.3	37.6
2,6-dimethylpyrazine	----	----	0.3	4.8	32.3
2,3-dimethylpyrazine	0.3	1.8	0.2	0.4	----
TOTAL (ppm)	13.1	5.7	19.9	0.74	2.2

* L = lysine, A = asparagine, G = glucose, F = fructose,
  R = ribose

The effect of type of amino acid on relative distribution of pyrazines can be seen in comparing lysine–sugar systems with asparagine–sugar systems. Relative yield (percentage of total pyrazines formed) of unsubstituted pyrazine is much greater for lysine than for asparagine with 56% for lysine–glucose versus 19% for asparagine–glucose, and 42% for lysine–fructose versus 1% for asparagine–fructose. For 2-methylpyrazine, no trend was apparent as relative yield was less for lysine–glucose (42%) than for asparagine–glucose (59%) but reversed with a greater relative yield for asparagine–fructose (49%) versus lysine–fructose (29%). Relative yields of the dimethylpyrazines were greater with asparagine versus lysine, with 22% for asparagine–glucose versus 3% for lysine–glucose and 70% for asparagine–fructose versus 9% for lysine–fructose. Under the conditions studied (95°C, 2 h treatment) no 2,6-dimethylpyrazine was detected in measurable quantities in the lysine systems, except with ribose, whereas in the asparagine systems this compound was detected, especially in considerable relative quantities in the asparagine–fructose system. The reverse situation occurred with 2,3-dimethylpyrazine. None was detected in the asparagine–fructose system and only 3 ppb in the asparagine–glucose system versus 30-100 ppb in the lysine systems.

The effect of type of sugar on relative distributions of pyrazines was most pronounced in the asparagine systems. For pyrazine, relative yield was greater with glucose versus fructose, with 56% for lysine–glucose versus 42% for lysine–fructose and 19% for asparagine–glucose versus 1% for asparagine–fructose. For 2-methylpyrazine, relative yields were close in the lysine systems, with 42% for lysine–glucose versus 49% for lysine–fructose. However, in the asparagine systems relative yields were greater with glucose versus fructose, with 59% for asparagine–glucose versus 29% for asparagine–fructose. For dimethylpyrazines, relative yields in both lysine and asparagine systems were always greater with fructose versus glucose, lysine–fructose with 9% versus lysine–glucose, 3%, and asparagine–fructose with 70% versus asparagine–glucose, 22%. This suggests that relative distributions of pyrazines formed is a function of sugar fragmentation.

The results from these studies compared favorably with the results of van Praag et al. (15) who found the same series of pyrazines resulted, regardless of source of sugar. Also, Koehler et al. (16) found that the source of carbohydrate affects the total yield of pyrazines in ammonium chloride–sugar systems. Yields were greater with fructose versus glucose. Koehler and Odell (22) also found that the ratio of dimethylpyrazine to to methylpyrazine differed, depending on sugar reacted.

Shibamoto and Bernhard (18) investigated the effect of sugar source on relative distributions of pyrazines. They found minor differences in relative distributions of pyrazines among the different sugars reacted with ammonium hydroxide. This appears to contradict the results of the current study. They concluded that since glucose is readily transformed to fructose and to a lesser amount of mannose through an enediol reaction intermediate in alkaline solutions, the pathway of pyrazines formation for these three sugars is similar which results in similar pyrazine distributions. The reaction systems which they used consisted of 8M ammonium hydroxide and 1M carbohydrate in water which results in reaction mixtures of extremely high basicity. pH has an effect both on the mutarotation of sugars and enolization/isomerization (32). Although Isbell (32) has demonstrated that D–glucose enolizes at a faster rate than D–fructose and Overend et al. (34) have found that percentages of acyclic conformation of sugars to vary, the extreme conditions in basicity may have acted to negate these differences, so that the net result was similar yields of pyrazines and their distributions. Conditions in the present study were milder at pH 9.0, so differing rates of mutarotation and enolization/isomerization of sugars may have allowed for the formation of different quantities of reaction intermediates, resulting in different total yields of pyrazines as well as distributions. Like the results of the current study, Shibamoto and Bernhard (18) did find pyrazine total yields were greater with pentoses than with hexoses. Unlike their study, the current study employed buffered solutions. This may have had an effect on pyrazine formation.

Burton and McWeeney (35) investigated the stability of sugars in the nonenzymatic browning reaction. They found that the development of chromophores proceeded at a faster rate for pentoses than hexoses and that the mutarotation velocities, as measured by polarographic wave heights, correlated with chromophore production. It seems reasonable to predict greater rates of pyrazine formation with increased rates of mutarotation of sugars. This was seen in the current study with systems containing a pentose (ribose) versus hexoses (glucose and fructose) in reaction with lysine. As far as predicting reactivities of glucose versus fructose, previous studies indicate that this is a bit more complex. Although Burton and McWeeney (35) found ketoses gave greater polarographic waves than the corresponding aldoses (fructose versus glucose), Kato et al. (36) found that fructose initially reacts faster than glucose as measured by color development, but then the values for glucose surpass fructose. This may be why total pyrazine yields were greater in the lysine-glucose system when reacted for 2 h at 95°C, while the reverse was found in asparagine systems.

With regard to the effect of type of amino acid, Newell et al. (13) found the same series of pyrazines to result regardless of the amino acid reacted. Under the conditions of the current study, 2,6-dimethylpyrazine was detected in the asparagine systems but not in the lysine systems except in reaction with ribose. The other pyrazines detected were similar for all reaction systems. The difference in results here is probably a result of limits of detection. Koehler et al. (16) observed formation of a similar series of pyrazines when reacting glucose with asparagine, glutamine, glutamic acid, aspartic acid and ammonium chloride. However, both total yields and relative product distribution of pyrazines differed as was observed in the current study. They proposed that the distribution of pyrazines produced is a function of the ease of nucleophilic attack of the amino acid on the carbonyls. The current data support this theory.

Koehler and Odell (22) found yields of pyrazines to be greater with asparagine than lysine by a factor of 16. This conflicts with results of the current study, in which yields of pyrazines are always greater with lysine than with asparagine. However, the reaction systems employed in the two studies were different. Koehler and Odell (22) reacted 0.1 mole each of glucose and an amino acid in 100 ml diethyleneglycol and 20 ml of water at 120°C for 24 h. Therefore, pH and water activity of the two systems differed which may have exerted some effect on the reactivities of the amino acids. Also, the unsubstituted pyrazine was not quantified in the previous study. This pyrazine was the major one produced in the current study. Although the ∝-amino group is most likely the major amino group reactive in the Strecker reaction, the Ɛ-amino group will be reactive in sugar-amino condensation, which will eventually result in greater yields of sugar fragmentation products. One would not expect the amide group of asparagine to participate in the Maillard reaction since it is a relatively stable neutral moiety. Therefore, greater yields of pyrazines with lysine versus asparagine seem quite reasonable.

Summary

In this investigation of the effects of types of sugars and amino acids on pyrazine formation, rate of formation best fit pseudo zero order reaction kinetics. Activation energies for pyrazine formation ranged from 27-45 kcal/mole. These values are higher than those reported in the literature for other aspects of the Maillard reaction suggesting a different rate-controlling step. Mean activation energies for dimethylpyrazine formation are slightly higher than those of the unsubstituted and methylpyrazine formation. Both type of amino acid and type of sugar had an effect on total yield and relative distributions of pyrazines.

Literature Cited

1. Maga, J.A., Sizer, C.E. J. Agric. Food Chem. 1973, 21, 22.

2. Maga, J.A., Sizer, C.E. In Fenaroli's Handbook of Flavor Ingredients; Furia, T.E., Bellanca, N., Eds.; CRC Press: Cleveland, 1975; Vol. 1, p. 47.

3. Maga, J.A. In Food Flavours. Part A. Introduction; Morton, I.D., MacLeod, A.J., Eds.; Elsevier Publishing Co.: Amsterdam. 1982; p. 283.

4. Maga, J.A. CRC Crit. Rev. Food Sci. Nutr. 1982, 16, 1.

5. Ohloff, G. and Flament, I. In Fortsch. Chem. Org. Naturst.; Herz, W., Grisebach, H., Kirby, G.W., Eds;, Springer Verlag: Vienna, 1979, p. 47.

6. Brophy, J.J., Cavill, G.W.K. Heterocycles 1980, 14, 477.

7. Barlin, G.B. The Chemistry of Heterocyclic Compounds; John Wiley & Sons; New York, 1982, Vol. 41.

8. Vernin, G. and Parkanyi, C. In The Chemistry of Heterocyclic Flavouring and Aroma Compounds; Vernin, G., Ed. Ellis Horwood, Ltd., Chichester, England, 1982; p. 120.

9. Vernin, G. and Verning, G. In The Chemistry of Heterocyclic Flavouring and Aroma Compounds; Vernin, G., Ed. Ellis Horwood, Ltd., Chichester, England, 1982; p. 120.

10. Shibamoto, T. In Instrumental Analysis of Foods; Charalambous, G., Inglett, G. Eds., Academic Press, New York; 1983; Vol. 1, p.229.

11. Hodge, J.E., J. Agric. Food Chem., 1953, 1, 928.

12. Dawes, I.W., Edwards, R.A., Chem. Ind., 1966, 2203.

13. Newell, J.A., Mason, M.E., Matlock, R.S., J. Agric. Food Chem. 1967, 15, 767.

14. Wang, P.S., Odell, G.V., J. Agric. Food Chem. 1973, 21, 868.

15. van Praag, M., Stein, H., Tibbetts, M., J. Agric. Food Chem. 1968, 16, 1005.

16. Koehler, P.E., Mason, M.E., Newell, J.A., J. Agric. Food Chem. 1969, 17, 393.

17. Rizzi, G.P. 1972. J. Agric. Food Chem. 1972, 20, 1081.
18. Shibamoto, T. and Bernhard, R.A. J. Agric. Food Chem. 1977, 25, 609.

19.   Shibamoto, T. and Bernhard, R.A. Agric. Biol. Chem. 1977, 41, 143.

20.   Wong, J., Bernhard, R. J. Agric. Food Chem. 1988, 36, 123.

21.   Rizzi, G. J. Agric. Food Chem. 1988, 36, 349.

22.   Koehler, P.E., Odell, G.V. J. Agric. Food Chem. 1970, 18, 895.

23.   Dwivedi, B.K. CRC Crit. Rev. Food Technol. 1975, 5, 487.

24.   Colowick, S.P., Kaplan, N.O., Eds. In Methods of Enzymology, Vol. I. Acad. Press, N.Y. 1955.

25.   Heller, S.R. and Milne, G.W.A. EPA/NIH Mass Spectral Data Base. Nat. Stand. Ref. Data Serv., Nat. Bur. Stand. (U.S.) U.S. Govt. Printing Office, Wash., D.C. 1978, 63.

26.   Labuza, T.P. 1981. In Applications of Computers in Food Research and Food Industry; Saguy, I., Ed.; Marcel Dekker: New York. 1981.

27.   Steele, R.G., Torrie, J.H. 1980. Principles and Procedures of Statistics; McGraw-Hill: New York. 1980.

28.   Labuza, T.P., Tannebaum, S.R., Karel, M. Food Technol. 1970 24, 35.

29.   Warmbier, H.C., Schnickels, R.A., Labuza, T.P. J. Food Sci. 1976, 41, 981.

30.   Lee, C.M., Sherr, B., Koh, Y-N. J. Agric. Food Chem. 1984, 32, 379.

31.   Stamp, J.A. and Labuza, T.P. J. Food Sci. 1983, 48, 543.

32.   Speck, J.C., Jr. Advan. Carbohyd. Chem. 1958, 13, 363.

33.   Isbell, H.S. In Carbohydrates in Solution; American Chemical Society: Washington, D.C., 1973; p. 70.

34.   Overend, W.G., Peacocke, A.R., Smith, J.B. J. Chem. Soc. 1961, 3487.

35.   Burton, H.S., McWeeney, D.J. Nature 1963, 197, 266.

36.   Kato, H., Yamamoto, M., Fujimaki, M. Agric. Biol. Chem. 1969, 33, 939.

RECEIVED September 23, 1988

Chapter 8

Formation and Aroma Characteristics of Heterocyclic Compounds in Foods

Chi-Tang Ho and James T. Carlin[1]

Department of Food Science, New Jersey Agricultural Experiment Station, Cook College, Rutgers University, New Brunswick, NJ 08903

Heterocyclic compounds have been identified as important volatile components of many foods. The odor strength and complexity of these compounds makes them desirable as flavoring ingredients.
   Heterocyclic compounds are primarily formed through nonenzymatic browning reactions. Recent studies of deep-fat fried food flavors led to the identification of pyrazines, pyridines, thiazole, oxazoles and cyclic polysulfides which had long-chain alkyl substitutions on the heterocyclic ring. The involvement of lipid or lipid decomposition products in the formation of these compounds could account for the long-chain alkyl substitutions.

Our knowledge of the chemical composition of food flavors has made considerable progress during the last twenty years. This is mainly due to advances in analytical techniques, such as the coupling of GC with MS and the development of fused-silica capillary columns. Heterocyclic compounds occupy a prominent position among the more than 10,000 compounds occurring in the volatiles of foods. This results from their exceptional sensory properties (1). Heterocyclic compounds contain one or more heteroatoms (O, S and/or N) in rings or fused ring systems.

   The majority of heterocyclic compounds are formed through thermal interactions of reducing sugars and amino acids, known as the Maillard reaction. Other thermal reactions such as hydrolytic and pyrolytic degradation of food components (e.g. sugars, amino acids, vitamins) and the oxidation of lipids also contribute to the formation of heterocyclic compounds responsible for the complex flavor of many foodstuffs. Heterocyclic compounds may also be formed enzymatically in vegetables (tomatoes, bell peppers, aspara-

[1]Current address: Joseph E. Seagrams & Sons, White Plains, NY 10604

0097–6156/89/0388–0092$06.00/0
○ 1989 American Chemical Society

gus), fruits (pineapple, passion fruit) and during the ripening of cheese.

Recent studies in our laboratory showed that lipids may be directly associated with the Maillard reaction in the formation of some heterocyclic compounds. The effect of lipids on the formation of heterocyclic compounds in a model Maillard reaction has also been reported by Mottram and Whitfield (2).

This paper discusses the formation and aroma characteristics of selected classes of heterocyclic compounds important to the flavor of foods, especially deep-fat fried foods.

Pyrazines

Alkylpyrazines have been recognized as important trace flavor components of a large number of cooked, roasted, toasted and deep-fat fried foods (3). As a rule, alkylpyrazines have a roasted nut-like odor and flavor. Formation pathways for alkylpyrazines have been proposed by numerous researchers (4, 5, 6). Model studies suggest that they are minor products of the Maillard reaction.

The recent identification of 2-heptylpyrazine in french fried potato flavor, and 2-methyl-3(or 6)-pentylpyrazine, 2-methyl-3(or 6)-hexylpyrazine and 2,5-dimethyl-3-pentylpyrazine in a heated and extruded corn-based model system, deserve special attention (7, 8, 9). These alkylpyrazines have a long-chain substitution on the pyrazine ring. Only the involvement of lipids or lipid-decomposition products in the formation of these compounds could account for the long-chain alkyl substitution on the pyrazine ring. A mechanism for the formation of 2,5-dimethyl-3-pentylpyrazine was proposed and is shown in Figure 1. 3,6-Dihydropyrazine, formed by the condensation of aminoketones, reacts with pentanal, a lipid oxidation product, and results in the formation of 2,5-dimethyl-3-pentylpyrazine. The possible reactivity of 3,6-dihydropyrazine with carbonyl compounds has been discussed by Flament (10). Rizzi (11) reported the formation of 2,5-dimethylpyrazine and 2,6-dimethylpyrazine in the reaction of 1-hydroxy-2-propanone (acetol) with ammonium acetate under mild conditions and acidic pH. The proposed mechanism was also supported by our identification of 2,5-dimethyl-3-pentylpyrazine and 2,6-dimethyl-3-pentylpyrazine as the major products when pentanal was added to a mixture of acetol and ammonium acetate and reacted at an elevated temperature (100°C). 2-Heptylpyrazine has green, waxy and earthy notes and could be an important contributor to the flavor of french fried potatoes or other fried food systems.

Various isopentyl-substituted pyrazines, such as 2-isopentyl-3-methylpyrazine, 2-isopentyl-5-methylpyrazine, 2-isopentyl-6-methylpyrazine, 2-isopentyl-5,6-dimethylpyrazine, 2-isopentyl-3,5-dimethylpyrazine and 2-isopentyl-3,6-dimethylpyrazine were identified from the thermal reaction of glucose and leucine (12). The formation mechanisms for these compounds may also involve the reaction of 3,6-dihydropyrazine with isovaleraldehyde, the Strecker aldehyde of leucine. Kitamura and Shibamoto (13) described 2-isopentyl-5,6-dimethylpyrazine as having a caramel-like, coffee and sweet aroma. Although isopentyl-substituted pyrazines have not yet been reported in cocoa, they could, if present, be very important contributors to that characteristic aroma.

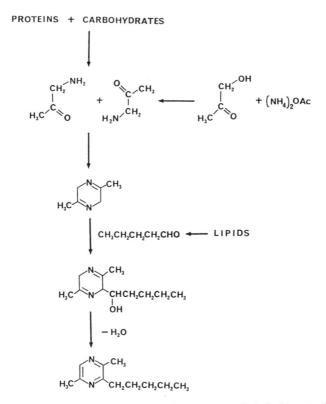

Figure 1. Mechanism for the formation of 2,5-dimethyl-3-pentylpyrazine.

Although most of the alkylpyrazines are formed through thermal interactions of components in food, methoxy-substituted pyrazines are mainly derived from biosynthetic pathways. 2-Isobutyl-3-methoxy-pyrazine isolated from bell pepper by Buttery et al. (14) is one of the most significant flavor compounds discovered. This characteristic bell pepper aroma compound has an extremely low odor threshold of 0.002 ppb in water (15).

Pyridines

The occurrence of pyridines in food has been reviewed (16). 2-Alkyl-pyridines were proposed to form from the corresponding unsaturated n-aldehydes with ammonia upon heat treatment (17, 18). Table I lists pyridines identified in the volatiles of fried chicken (19) and french fried potatoes (7).

Table I. Pyridines Identified in Fried Chicken and
French Fried Potato Flavor

Compound	Fried Chicken	French-Fried Potato
Pyridine	+	
2-Methylpyridine	+	
2-Ethylpyridine		+
3-Ethylpyridine	+	
4-Ethylpyridine	+	
2-Acetylpyridine		+
2-Methyl-5-ethylpyridine	+	
2-Ethyl-3-methylpyridine	+	
2-Butylpyridine	+	+
2-Pentylpyridine	+	+
2-Heptylpyridine		+
2-Pentyl-3,5-dibutylpyridine		+
2-Isobutyl-3,5-dipropylpyridine	+	

2-Pentylpyridine was identified in both fried chicken and french fried potato flavors. This compound has a strong fatty and tallow-like odor and was the major product in the volatiles generated from the thermal interaction of valine and linoleate (20). It is postulated to form through the reaction of 2,4-decadienal and ammonia (20). The reaction of 2,4-decadienal with either cysteine or glutathione (γ-glu-cys-gly) in aqueous solution at high temperature (180°C) yielded 2-pentylpyridine as the major product (Zhang, Y. and Ho, C.-T., Rutgers University, unpublished data). The amount of 2-pentyl-pyridine generated in the 2,4-decadienal/glutathione system did not differ significantly from that in the 2,4-decadienal/cysteine system. It is possible that the amino group from amino acids, or peptide, condenses directly with the aldehydic group of 2,4-decadienal and is then followed by an electrocyclic reaction and aromatization to form 2-pentylpyridine (Figure 2).
2-Isobutyl-3,5-diisopropylpyridine was identified in fried chicken and has a roasted cocoa-like aroma (21). Figure 3 shows the mechanism for the formation of this compound as proposed by Shu et

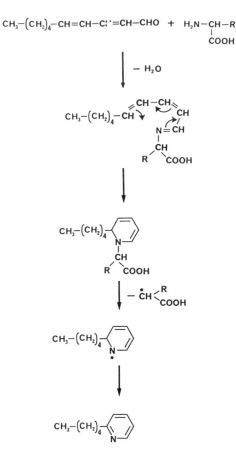

Figure 2.   Mechanism for the formation of 2-pentylpyridine.

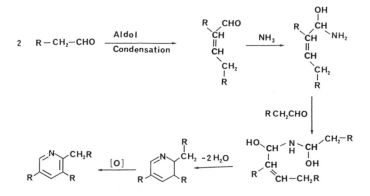

Figure 3.   Mechanism for the formation of 2,3,5-trialkyl-pyridine.

al. (22). It involves the reaction of aldehyde and ammonia at high temperatures and is known as the Chichibabin condensation. 2-Isobutyl-3,5-diisopropylpyridine has also been identified in a glucose/leucine model system possessing a cocoa-like aroma (23). 2-Pentyl-3,5-dibutylpyridine was identified in french fried potato flavor and presumably formed through a Chichibabin reaction involving ammonia and hexanal, an abundant thermal oxidative decomposition product of lipids.

Thiazoles

Thiazoles are a class of compounds possessing a five-membered ring with sulfur and nitrogen in the 1 and 3 positions, respectively. The potential for thiazole derivatives as flavorants is evident from the work of Stoll et al. (24) who found the strong nut-like odor of a cocoa extract to be due to a trace amount of 4-methyl-5-vinylthiazole. Since then, numerous thiazoles have been identified in food flavors.

The exact origin of thiazoles remains a mystery. They might form through the thermal degradation of cystine or cysteine (25, 26), or by the interaction of sulfur-containing amino acids and carbonyl compounds (27, 28). Thiazoles have been identified as volatile components of thermally degraded thiamine (29).

Aroma properties of some alkylthiazoles have been reviewed (30). The most typical alkylthiazole is probably 2-isobutylthiazole. This compound was isolated from tomato flavor and was described as having a strong green odor resembling that of tomato leaf (31). When added to canned tomato puree or paste at levels of 20 to 50 ppb, 2-isobutylthiazole develops an intense fresh tomato-like flavor. Most of the alkylthiazoles are described as green, nutty and vegetable-like (32, 33).

Table II lists alkylthiazoles identified in fried chicken flavor (19) and french fried potato flavor (7).

Several of the alkylthiazoles identified in french fried potato flavor, such as 2,4-dimethyl-5-propylthiazole, 2,4-dimethyl-5-pentylthiazole, 2-butyl-4-methyl-5-ethylthiazole and 2-butyl-4-propylthiazole have a strong characteristic sweet, sulfury and green aroma (33). This aroma characteristic is quite distinctive and is present in a large number of the fractions generated from the gas chromatographic fractionation of the french fried potato flavor isolate. It is probably an important part of the total french fried potato flavor.

2-Pentyl-4-methyl-5-ethylthiazole has a strong paprika pepper flavor and 2-heptyl-4,5-dimethylthiazole has a strong spicy flavor. 2-Octyl-4,5-dimethylthiazole has a sweet fatty aroma (33). They are probably important contributors to the flavor of fried foods. These thiazoles and other thiazoles identified have long-chain alkyl substitutions on the thiazole ring. The involvement of frying fat or fat decomposition products in the formation of these compounds is again suggested.

Thiazolines, a reduced form of thiazoles, have also been reported to occur in foods, mainly cooked beef. 2,4,5-Trimethyl-2-thiazoline was identified in beef broth (34). 2,4-Dimethyl-3-thiazoline found in cooked beef aroma was reported to have a nutty, roasted and vegetable aroma (35).

Table II. Alkylthiazoles Identified in Fried Chicken
and French Fried Potato Flavor

Compound	Fried Chicken	French-Fried Potato
Thiazole	+	
2-Methylthiazole	+	
2,4,5-Trimethylthiazole	+	+
2-Methyl-4-ethylthiazole	+	
2,4-Dimethyl-4-ethylthiazole	+	
2-Ethyl-4,5-dimethylthiazole		+
2-Isopropyl-4-methylthiazole		+
2-Propyl-4,5-dimethylthiazole	+	+
2-Isopropyl-4,5-dimethylthiazole		+
2,5-Dimethyl-4-butylthiazole	+	+
2-Isopropyl-4-ethyl-5-methylthiazole	+	+
2-Butyl-4,5-dimethylthiazole	+	+
2-Isobutyl-4,5-dimethylthiazole		+
2,4-Dimethyl-5-pentylthiazole		+
2,4-Diethyl-5-propylthiazole		+
2-Butyl-4-methyl-5-ethylthiazole	+	+
2-Butyl-4-propylthiazole		+
2-Pentyl-4,5-dimethylthiazole	+	+
2-Butyl-4-propyl-5-methylthiazole		+
2-Pentyl-4-methyl-5-ethylthiazole		+
2-Pentyl-5-propylthiazole		+
2-Hexyl-4,5-dimethylthiazole	+	+
2-Heptyl-4,5-dimethylthiazole	+	+
2-Heptyl-4-ethyl-5-methylthiazole	+	+
2-Octyl-4,5-dimethylthiazole	+	+
2-Octyl-4-methyl-5-ethylthiazole		+

Oxazoles

Oxazoles are characterized by possessing a five-membered ring with
oxygen and nitrogen in the 1 and 3 positions. The occurrence of
oxazoles in food flavor has been reviewed (36). Recently twenty-
four alkyloxazoles were identified in the volatile compounds from
french-fried potaotes (37). This represents the largest number of
oxazoles reported in a food system. Like the alkylthiazoles, several of the alkyloxazoles identified, such as 2-pentyl-4,5-dimethyl-
oxazole, 2-pentyl-4-methyl-5-ethyloxazole, 2-hexyl-4,5-dimethyloxa-
zole and 2-hexyl-4-methyl-5-ethyloxazole, have long-chain alkyl substitutions at the 2-position of the oxazole ring. Lipids or lipid
decomposition products could be involved in the formation of these
long-chain alkyl substituted oxazoles.

Oxazoles have a wide range of aroma characteristics. As an example, Table III lists the aroma characteristics of some alkyloxazoles identified in fried bacon flavor (Lee, K. N., Rutgers University, unpublished data).

Table III. Odor Description of Alkyloxazoles
Identified in Fried Bacon Flavor

Oxazole	Odor Description
5-butyloxazole	very strong, bacon fatty, Animalic, aged meat-like
2,5-dimethyl-5-propyloxazole	green, flowery, sweet
2-methyl-5-pentyloxazole	sweet, strong floral, fatty-waxy
2,5-dimethyl-4-butyloxazole	fresh acidic green, pickle-like
2,5-dimethyl-5-butyloxazole	dry herbal, seasonal herbal
2-isopropyl-4-ethyl-5-methyloxazole	sweet, fruity
2-phenyl-5-ethyloxazole	indole-like, phenolic

It is interesting to note that 5-butyloxazole has a very distinct bacon-fatty aroma and could be an important flavor constituent of fried bacon. 5-Pentyloxazole also possesses a similar aroma characteristic. Both 5-butyloxazole and 5-pentyloxazole have no alkyl group on carbon 2 or 4 of the oxazole ring. When a methyl group is substituted on carbon-2 (e.g., 2-methyl-5-pentyloxazole), the fatty aroma decreases and a sweet-floral aroma becomes more characteristic. The sweet-floral character is further enhanced by additional methyl substitution on carbon-4 (38).

2-Pentyl-4-methyl-5-ethyloxazole was identified in french-fried potato flavor and has a strong buttery, sweet and lactone-like flavor. It is probably an important contributor to the fried food aspect of french-fried potato flavor (37).

Oxazoles and thiazoles possessing comparable alkyl groups were reported to have significant aroma similarities (33). Buttery et al. reported that some 4,5-dialkylthiazoles possessed potent bell pepper aroma (39). The most potent one, 4-butyl-5-propylthiazole, was reported to have a flavor threshold of 0.003 ppb in water. Ho and Tuorto (40) synthesized several 4,5-dialkyloxazoles and found them to have a green, vegetable-like aroma. 4-Butyl-3-propyloxazole has a strong bell pepper aroma and a flavor threshold of 0.1 ppm in water. 5-Pentylthiazole also had strong fatty and sweet aromas reminiscent of 5-pentyloxazole. 2-Pentyl-5-methylthiazole was judged to have a fermented vegetable-like aroma. The corresponding 2-pentyl-5-methyloxazole was described as acidic and sweet with a flowery afternote (38).

Figure 4 shows a proposed mechanism for the formation of 2,4,5-trimethyloxazole and 4,5-dimethyloxazole from the Strecker degradation of cysteine with 2,3-butanedione (41).

3-Oxazolines, the reduced form of oxazoles, also have important sensory properties. The first report of a 3-oxazoline was made by

Chang et al. (42). They isolated and identified 2,4,5-trimethyl-3-
oxazoline in boiled beef. This compound was described as having a
"characteristic boiled beef aroma". Mussinan et al. (35) identified
oxazolines and no oxazoles in their beef system. Peterson et al.
(43) reported on the volatiles of canned beef stew. Both 2,4,5-
trimethyloxazole and 2,4,5-trimethyl-3-oxazoline were present. The
relative concentration of 2,4,5-trimethyloxazole was medium while
for 2,4,5-trimethyl-3-oxazoline was extra high. Lee et al. (44)
identified 2-methyl-3-oxazoline, 2,4-dimethyl-3-oxazoline and 2,4,5-
trimethyl-3-oxazoline in the volatiles of roasted peanuts. The latter two 3-oxazolines were also identified in the volatiles of fried
chicken (19).

Trithiolanes

Trithiolanes have received increasing attention since the identification of diastereomeric 3,5-dimethyl-1,2,4-trithiolane in the volatiles of boiled beef (42). The parent 1,2,4-trithiolane is a component of Shiitake mushrooms (45) and red algae (27). In addition
to 3,5-dimethyl-1,2,4-trithiolane, Kubota et al. (46) identified 3-
methyl-5-ethyl-1,2,4-trithiolane and 3,5-diethyl-1,2,4-trithiolane
in both cis and trans forms in boiled Antarctic Gulls. Both compounds were described as garlicky. Flament et al. (47) reproted the
identification of 3-methyl-5-ethyl-1,2,4-trithiolane and 3-methyl-5-
isopropyl-1,2,4-trithiolane in a commercial beef extract.
    3,5-Diisobutyl-1,2,4-trithiolane was identified in the vola-
tiles isolated from fried chicken (21). This compound has been reported to possess roasted, roasted-nut, crisp bacon-like and pork
rind-like aromas and flavors (48). In addition to 3,5-dimethyl-1,2,
4-trithiolane and 3,5-diisobutyl-1,2,4-trithiolane, two long-chain
alkyl substituted trithiolanes, namely, 3-methyl-5-butyl-1,2,4-tri-
thiolane and 3-methyl-5-pentyl-1,2,4-trithiolane, were reported to
be present in fried chicken flavor (49). Along with 3,5-dimethyl-
1,2,4-trithiolane, 3-methyl-5-ethyl-1,2,4-trithiolane, 3-methyl-5-
propyl-1,2,4-trithiolane and 3-methyl-5-butyl-1,2,4-trithiolane are
reported to be important flavor components of Chinese stewed pork
(50).
    A mechanism has been reported for the formation of trithiolane
from the reaction of aldehydes with hydorgen sulfide (51). The
identification of 3-methyl-5-butyl-1,2,4-trithiolane and 3-methyl-5-
pentyl-1,2,4-trithiolane in food flavor suggests that pentanal and
hexanal were involved in the formation of these compounds (Figure
5). Pentanal and hexanal are major thermal and oxidative decomposition products of lipids.

Summary

Heterocyclic compounds, especially those which contain nitrogen and
sulfur atoms, possess potent sensory qualities at low concentrations. They are formed in foods by thermal decomposition and interaction of food components. The identification of many long-chain
alkyl substituted heterocyclic compounds suggests that their formation mechanisms directly involve lipids or lipid decomposition
products.

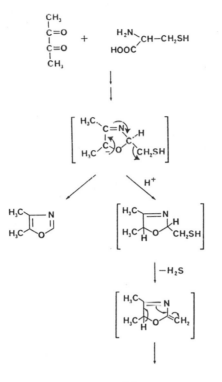

Figure 4. Mechanism for the formation of 2,4,5-trimethyl-oxazole.

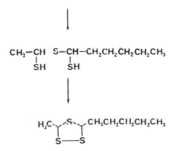

Figure 5. Mechanism for the formation of 3-methyl-5-pentyl-1,2,4-trithiolane.

Acknowledgments

New Jersey Agricultural Experiment Station Publication No. D-10205-3-88 supported by State Funds and Regional Project NE-116. We thank Mrs. Joan B. Shumsky for her secretarial help.

Literature Cited

1. Garnero, J. In The Chemistry of Heterocyclic Flavoring and Aroma Compounds; Vernin, G. Ed.; John Wiley & Sons: New York, 1982, p. 17.
2. Mottram, D. S.; Whitfield, F. B. In Flavour Science and Technology; Martens, M.; Dalen, G. A.; Russwurm, H., Jr. Ed.; John Wiley & Sons: New York, 1987, p. 29
3. Maga, J. A. CRC Crit,. Rev. Food Sci. Nutr. 1982, 16, 1-115.
4. Rizzi, G. P. J. Agric. Food Chem. 1972, 20, 1081-1085.
5. Shibamoto, R.; Bernhard, R. A. J. Agric. Food Chem. 1977, 25, 609-614.
6. Wong, J. M.; Bernhard, R. A. J. Agric. Food Chem. 1988, 36, 123-129.
7. Carlin, J. T. Ph.D. Thesis, Rutgers University, New Jersey, 1983.
8. Huang, T.-C.; Bruechert, L. J.; Hartman, T. G.; Rosen, R. T.; Ho, C.-T. J. Agric. Food Chem. 1987, 35, 985-990.
9. Bruechert, L. J. M.S. Thesis, Rutgers University, New Jersey, 1987.
10. Flament, I. In The Quality of Foods and Beverages; Charalambous, G.; Inglett, G., Ed.; Academic: New York, 1981, p. 35.
11. Rizzi, G. P. J. Agric. Food Chem. 1988, 36, 349-352.
12. Hwang, S. S. Ph.D. Thesis, Rutgers University, New Jersey, 1986.
13. Kitamura, K.; Shibamoto, T. J. Agric. Food Chem. 1981, 29, 188-192.
14. Buttery, R. G.; Seifert, R. M., Lundin, R. E.; Guadagni, D. G.; Ling, L. C. Chem. Ind. 1969, 490-491.
15. Seifert, R. M.; Buttery, R. G.; Guadagni, D. G.; Black, D. R.; Harris, J. G. J. Agric. Food Chem. 1970, 18, 246-249.
16. Vernin, G.; Vernin, G. In The Chemistry of Heterocyclic Flavoring and Aroma Compounds; Vernin, G. Ed.; John Wiley & Sons: New York, 1982, p. 72.
17. Buttery, R. G.; Ling, L. C.; Teranishi, R.; Mon, T. R. J. Agric. Food Chem. 1977, 25, 1227-1230.
18. Ohnishi, S.; Shibamoto, T., J. Agric. Food Chem. 1984, 32, 987-992.
19. Tang, J.; Jin, Q. Z.; Shen, G.-H.; Ho, C.-T.; Chang, S. S. J. Agric. Food Chem. 1983, 31 1287-1292.
20. Henderson, S. K.; Nawar, W. W. J. Amer. Oil Chem. Soc. 1981, 58, 632-635.
21. Hartman, G. J.; Carlin, J. T.; Hwang, S. S.; Bao, Y.; Tang, J.; Ho, C.-T. J. Food Sci. 1984, 49, 1398 & 1400.
22. Shu, C.-K.; Mookherjee, B. D.; Bondarovich, H. A.; Hagdorn, M. L. J. Agric. Food Chem. 1985, 33, 130-132.

23. Hartman, G. J.; Scheide, J. D.; Ho, C.-T. Perfumer & Flavorist 1983, 8(6), 81-86.
24. Stoll, M.; Dietrich, P.; Sundte, E.; Winter, M. Helv. Chim. Acta 1967, 50, 2065.
25. Shu, C.-K.; Hagedorn, M. L.; Mookherjee, B. D.; Ho, C.-T. J. Agric. Food Chem. 1985, 33, 438-442.
26. Shu, C.K.; Hagedorn, M. L.; Mookherjee, B. D.; Ho, C.-T. J. Agric. Food Chem. 1985, 33, 442-446.
27. Ohloff, G.; Flament, I. Fortschr. Chem. Org. Naturst 1978, 36, 231-283.
28. Hartman, G. J.; Ho, C.-T. Lebensm.-Wiss. u. -Technol. 1904, 17, 171-174.
29. Hartman, G. J.; Carlin, J. T.; Scheide, J. D.; Ho, C.-T. J. Agric. Food Chem. 1984, 32, 1015-1018.
30. Fors, S. ACS Symp. Ser. 1983, 215, 183-286.
31. Viani, R.; Bricout, J.; Marion,J. P.; Muggler-Chavan, F.; Reymond, D.; Egli, R. H. Helv. Chim. Acta 1969, 52, 887-891.
32. Pittet, A.O.; Hruza, D. E. J. Agric. Food Chem. 1974, 22, 264-269.
33. Ho, C.-T.; Jin, Q. Z. Perfumer & Flavorist 1984, 9(6), 15-18.
34. Tonsbeek, C. H.T.; Copier, H.; Plancken, A. J. J. Agric. Food Chem. 1971, 19, 1014-1016.
35. Mussinan, C. J.; Wilson, R. A.; Katz, T.; Hruza, A.; Vock, M. H. ACS Symp. Ser. 1976, 26, 133-145.
36. Maga, J. A. Crit. Rev. Food Sci. Nutr. 1981, 12, 295-307.
37. Carlin, J. T.; Jin, Q. Z.; Huang, T.-Z.; Ho, C.-T.; Chang, S. S. J. Agric. Food Chem. 1986, 34, 621-623.
38. Jin, Q. Z.; Hartman, G. J.; Ho, C.-T. Perfumer & Flavorist 1984, 9(4), 25-29.
39. Buttery, R. G.; Guadagni, D.; Lundin, R. J. Agric. Food Chem. 1976, 24, 1-6.
40. Ho, C.-T.; Tuorto, R. M. J. Agric. Food Chem. 1981, 29, 1306-1308.
41. Ho, C.-T.; Hartman, G.J. J. Agric. Food Chem. 1982, 30, 793-794.
42. Chang, S. s.; Hirai, C.; Reddy, B. R.; Herz, K. O.; Kato, A.; Sipma, G. Chem. Ind. 1968, 1639-1640.
43. Peterson, R. J.; Izzo, H. J.; Jungermann, E.; Chang, S. S. J. Food Sci. 1975, 40, 948-954.
44. Lee, M.-H.; Ho, C.-T.; Chang, S. s. J. Agric. Food Chem. 1981, 29, 684-686.
45. Chen, C.-C.; Ho, C.-T. J. Agric. Food Chem. 1986. 34, 830-833.
46. Kubota, K.; Kobayashi, A.; Yamanishi, T. Agric. Biol Chem. 1980, 44, 2677.
47. Flament, I.; Willhalm, B.; Ohloff, G. In Flavor of Foods and Beverages; Charalambous, G.; Inglett, G., Eds.; Academic: New York, 1978, p. 15.
48. Shu, C.-K.; Mookherjee, B. D.; Vock, M. H. U. S. Patent 4,263,331, 1981.
49. Hwang, S.-S.; Carlin, J. T.; Bao, Y.; Hartman, G. J. J. Agric. Food Chem. 1986, 34, 538-542.

50. Chou, C.C.; Wu, C.-M. FIRDI Research Report No. 285, 1983.
51. Takken, H. J.; Van der lInde, L.M.; De Valois, P. J.; Van Dort,
     H. M.; Boelen, M. ACS Symp. Ser. 1976, 26, 114-121.

RECEIVED August 5, 1988

Chapter 9

Natural Flavors Produced by Biotechnological Processing

David W. Armstrong[1], Bruce Gillies[2], and Hiroshi Yamazaki[2]

[1]Division of Biological Sciences, National Research Council of Canada,
Ottawa K1A 0R6, Canada
[2]Institute of Biochemistry, Carleton University, Ottawa K1S 5B6, Canada

The world market for flavors is growing
steadily with a distinct trend towards
'natural' compounds. Biotechnology offers
many advantages over traditional
extraction of botanical materials for
flavors production. These include highly
specific end product generation (e.g.
optically-active compounds), high yields
and purity, along with guaranteed supply.
Commercial exploitation of biotechnology
in this area not only relies upon
technical advances but as well on satisfy-
ing certain regulatory considerations.
These aspects are highlighted using
examples of whole cell microbial culture
and isolated enzyme systems.

A recent Japanese report predicts that products from the
food and beverage industry will top the list of
biotechnology products in sales by the year 2000 (1).
While demand for food products can fluctuate,
particularly at the bulk end of the market, certain high
priced products such as flavor/aroma compounds are
experiencing a constant increase in demand (2).
Flavoring compounds, substances gratifying taste and
smell, represent 10-15% by weight of world-use food
additives, which amounts to 25% of the value of the
total food additives market (3). Recently, it has been
estimated that 5,000-10,000 natural flavor compounds
exist (4) with 4,300 having been identified (5).
    The market for flavors is anticipated to grow at an
annual rate of about 30% (6). An aging population and
the associated diminuation of taste acuity among older
persons has resulted in a need for products with more
intense flavors (7). Other market demands for a greater

0097–6156/89/0388–0105$06.00/0
Published 1989 American Chemical Society

variety of flavored products has also increased flavor
material consumption. A recent 'sensory audit' conducted
by a leading food processor revealed that consumers were
most concerned about 'freshness' and 'naturalness' in
products they buy (7). This trend can be ascribed to
increasing health and nutrition conscious life-styles
which has encouraged the development of natural food
products (8). One way food processors have responded to
this trend towards 'naturalness' is an increased use of
natural food flavorants found in plants (9).
Unfortunately, these raw materials are subject to
various problems (8). Often, the desired flavor
component is in low concentration and therefore
extraction from botanical material is expensive. As
well, the supply of the materials is subject to seasonal
variation and to the vagaries of the weather, which
significantly affects the yield and quality of the
flavor. The supply is also influenced by the
socio-political stability of the producing regions. The
industry, if it is to meet consumer demand effectively,
needs competitive new sources of natural flavors, and
biotechnology has become a necessary solution.

NATURAL FLAVORS. The meaning of the term 'natural' may
vary between different social communities. In the
United States the Code of Federal Regulations (CFR
101.22.a.3) defines the term 'natural flavor' to include
not only animal or plant derivatives but also includes
enzymic and fermentative processes. A natural flavor,
according to the U.S. FDA guidelines set down in 1958,
must be produced from natural starting materials and
that the end-product must be identical to something
already known to exist in nature. Thus, biocatalytic,
but not chemical, transformation of natural substances
can be legally labelled as natural.

Application of Biotechnological Processing for Natural
Flavors-Production

Biotechnological approaches for production of flavors
have the potential to alleviate the problems outlined
with traditional botanical-derived processing routes.
Additionally, other benefits can be expected which
should serve to control costs, supply and consistency.
    Food technologists have had much experience with
the use of industrial microorganisms and enzymes in
traditional processes (bread and cheese making, alcohol
fermentation). Numerous early reports, some dating back
to the early 1900s, realized the potential of
microorganisms for flavour/aroma production. However a
limited knowledge base prevented the application of many
biotechnological routes for flavor production. More
recently, the level of knowledge pertaining to metabolic
pathways and their regulatory mechanisms has allowed for

more control through physiological and/or genetic
manipulations. With rapid advances occurring in genetic
engineering and fermentation technology, the area of
biotechnological flavors production is sure to benefit.
There has been much work conducted recently in the
area of plant cell culture, or phytoproduction,
especially where the product is a plant-unique mixture
of individual flavor substances such as vanilla extract
of which vanillin is the major component. As well, the
possibility of genetically engineering improved
varieties of plants for high yield and consistent
quality products is of considerable interest especially
for more complex plant-unique flavors. Many flavor
compounds are secondary metabolites for which a detailed
understanding of their production is not well
understood. Presently more knowledge exists in
microbial metabolism relative to plant biosynthetic
pathways and therefore has resulted in more successful
development of microbial-based flavor bioprocessing. As
well, scale-up of microbial cultures and isolated
enzymes has become relatively common practice while the
translation of plant cell culture to large commercial
scales is not yet well established. This review will
focus on the microbial whole cell and isolated enzyme
systems for flavor production.

MICROBIAL WHOLE CELL AND ENZYMIC SYSTEMS: A number of
single flavour substances have been identified which are
associated with particular flavors (10):
2-isobutylthiazole with tomato flavor; methyl- and
ethyl-cinnamates with strawberry flavor; methyl
anthranilate with grape flavor; and benzaldehyde with
cherry flavor. Single flavor components such as these
should be amenable to microbial production from
nutrients via multienzyme steps or enzymic production
from appropriate precursors.
An important attribute of microbial whole cell
biocatalysts is the ability to synthesize products de
novo from relatively inexpensive nutrients such as
simple sugars. Citric acid, produced by the fungus
Aspergillus niger, is perhaps the best known flavor
compound. Another fungus Trichoderma reesei can produce
a strong coconut-like impression which is the result of
the formation of 6-pentyl-α-pyrone (chemical synthesis
requires seven steps). The yeast Sporobolomyces odorus
can produce the lactones 4-decanolide and
cis-6-dodecen-4-olide, responsible for a peach-like
impression (11). Apart from 'fruity' flavors there
seems to be a growing interest in the production of
natural flavor chemicals related to "toasted" or cooked
meat flavors for application to microwave products (12).
It has been shown that a mixture of certain sugars and
dihydroxyacetone (derived from a fermentation) give an

improvement in browning and taste due to the development
of roasting and aromatic compounds (13).
    Recently we have found that the yeast Candida
utilis can convert glucose, and other fermentable
sugars, to ethyl acetate (14). Ethanol can also be
converted directly to ethyl acetate or acetaldehyde by
this yeast (15) depending upon the concentration of
ethanol provided in the medium (Figure 1). Ethyl
acetate is important for its use in 'fruity' flavors
which acetaldehyde has an important application in
providing 'freshness' to many products including
fruit-based drinks (16). Both conversions operate when
iron limiting conditions are imposed and they
demonstrate that relatively simple physiological
manipulations can result in marked changes in
metabolism. This yeast is capable of 'aerobic'
fermentation and readily forms ethanol from fermentable
sugars. Under iron-sufficient aerobic conditions C.
utilis produces biomass and is currently used
commercially for this purpose (single cell protein).
The imposition of iron-limited conditions results in the
tricarboxylic acid (TCA) cycle being severely rate
limited (Figure 2). Certain iron-requiring enzymes in
the TCA cycle including succinate dehydrogenase and
aconitase become activity limited. Under
iron-limitation the intracellular level of acetyl-CoA
accumulates which in the presence of ethanol allows for
increased ethyl acetate formation via an alcolol
transferase mechanism. Through the same metabolic
scheme, when the level of ethanol is elevated above
approximately 3.5% (w/v), acetaldehyde accumulates. As
the rate of oxidation of ethanol to acetaldehyde exceeds
that of the oxidation of acetaldehyde to acetic acid,
the aldehyde begins to accumulate. This intracellular
accumulation forward inhibits acetyl-CoA synthetase
which results in a reduction in ethyl acetate formation
and a change in product distribution to acetaldehyde.
Apart from producing two important flavor substances,
the spent yeast itself can be used post-processing in
natural flavour applications. Candida utilis can be
used for imparting 'savory' or 'meaty' flavors as it is
classified by the U.S. FDA as GRAS
(Generally-Recognized-As-Safe) and is one of only three
yeasts allowable for this purpose in human food products
(17).
    Another important attribute of whole cell/enzyme
systems is for the production of optically-active
compounds. As early as the 1960s, the synthesis of
optically active gamma- and delta-lactones by
microbiological reduction was demonstrated (18).
Production of optically-pure L-glutamate for use in the
flavor enhancer MSG by microbial means is another
testimonial to this potential of biotechnology.

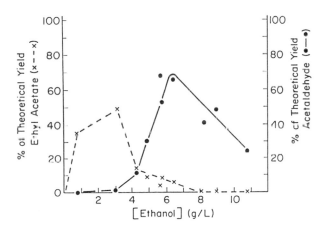

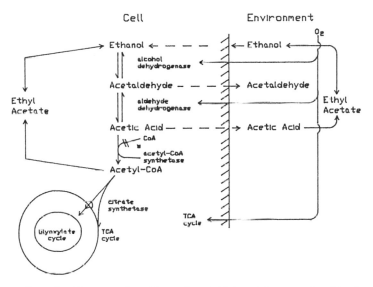

Fig. 1.   Change in product distribution from ethyl
acetate to acetaldehyde of <u>Candida</u> <u>utilis</u> due to
increasing levels of added ethanol.

Fig. 2.   Pathway of ethanol utilization and ethyl
acetate or acetaldehyde production by <u>Candida</u> <u>utilis</u>
TCA cycle activity inhibited under iron-limited
conditions.   Acetyl-CoA synthetase forward inhibited
by acetaldehyde accumulated when elevated levels of
ethanol ($\geq$ 3.5% w/v) present in the medium.

Transformation of precursor compounds to flavors
can also readily take advantage of biotechnological
processing. For example, the sesquiterpene valencene,
readily avaiable from orange oil but of low commercial
value, can be transformed by certain bacteria (18) to
the higher value sesquiterpene, nootkatone (grapefruit
citrus flavor) currently valued at about $4200/kg (19)
up from $1300-1600 a year earlier (synthetic $700/kg).
Another biotransformation having significant impact on
the citrus juice industry is the ability of A. niger to
convert the bitter component naringin to the non-bitter
form naringenin (20). Recent consumer demand for
grapefruit-based flavors is driving the price up for
naringin (currently priced at $55/kg and rising) which
could be converted biologically to an even higher priced
non-bitter form. An area where biotransformations would
improve product yields, and simultaneously result in an
optically-active end product, involves L-menthol
production. Currently the market for this terpene is
approximately 3000 tons/y and relies upon traditional
extraction of plant materials. In the peppermint plant,
oil 'maturation' occurs between the initial flowering
and full bloom, during which time only a small portion
(40%) of L-menthone is converted to L-menthol. It has
been proposed that L-menthone could be extracted at peak
levels from the plant and subsequently converted via
dehydrogenase activity of various microorganisms.
Various systems employing the bacterium Pseudomonas
putida and the yeast Rhodotorula minuta have resulted in
L-menthol with 100% optical activity (20). The
production of flavor complexes where possibly hundreds
of individual flavor substances contribute to an overall
flavor impression is highly suited to biotransformation.
To create the same complex flavor by the addition of
individual flavor components would be very difficult if
not impossible. As a result, the exogenous addition of
whole cells and/or enzymes to accelerate the flavor
maturation process of a number of food products is
rapidly attaining commercial reality. Cheese ripening
or aging is an area of intense research and should lead
to significant cost reductions. In the U.S. alone over
1 billion kilograms of Cheddar-type cheese is produced
annually. Normally, the aging process takes place over
a period of between 3-9 months at a cost of 1-2
cents/kg.month. The aging process involves primarily
the action of endogenous biocatalysts which transforms a
relatively bland elastic mass to a well-bodied cheddar
cheese. Research has shown that two key enzymic
mechanisms involving lipase and protease are critical to
cheese aging (21). The lipase preferentially hydrolyses
triglycerides, yielding C_6-C_{10} free fatty acids while
the protease provides a balanced flavor development
through proteolytic cleavage. A number of products are
now available commercially for accelerated cheese

ripening including a lipase/protease preparation derived
from <u>Aspergillus</u> <u>oryzae</u>. Lipases from <u>Candida</u>
<u>cylindracae</u> (rugosa) and <u>Mucor</u> <u>miehei</u> show different
specificity for fatty acid chain lengths and have been
used to generate different cheese flavor complexes.
  Apart from producing flavor substances or
complexes, biocatalysts could facilitate current
extraction processes. A Japanese process uses fungal
cellulase enzymes to enhance juice extraction from
oranges without disrupting the rind oil glands which
contain 'off' flavors and bitterness. Cellulases and
pectinases have also been used to degrade plant cell
walls to release more oil from oil seeds during
processing. The use of 500-1000 g of enzyme per ton of
seeds at temperatures of 30-50°C can produce a 2-6%
increase in oil yield (<u>22</u>).

NOVEL FLAVOR BIOPROCESSING SYSTEMS. Typically the
systems described above are operated in aqueous
environments. It has recently become evident that many
enzymes (<u>23</u>) and even certain whole cells (<u>24</u>) can
function in apolar or organic solvents such as hexane.
Although the above systems are referred to as
'non-aqueous', technically, the enzymes have to be in a
biphasic system (organic solvent/water) with the aqueous
component being present in low amounts, even down to a
monomolecular layer on an enzyme. This knowledge will
certainly benefit biotechnological production of flavor
production for a number of reasons. Since many flavor
compounds or their precursors have limited water
solubility, the use of apolar solvents rather than water
could allow for more efficient conversions due to better
enzyme/substrate interaction. In addition, some
reactions may only occur to a significant degree in a
more apolar environment such as esterification with
lipase enzymes (<u>25</u>). Normally, in the case of a
lipase-based conversion, esters are hydrolytically
converted to the acid and alcohol moieties. In certain
apolar solvents this reaction can be reversed owing to
the preferred partitioning of the ester away from the
enzyme thereby 'pulling' the reaction (Figure 3). More
detailed discussions on the use of organic media for
biocatalysis can be found elsewhere (<u>26</u>).
  We have conducted studies related to the production
of flavor esters using immobilized lipase from the yeast
<u>C</u>. <u>cylindracae</u> in a non-aqueous system (<u>25</u>, <u>27</u>). The
lipase from this microorganism was found to have a broad
range specificity and was useful to produce a number of
commercially important esters including ethyl butyrate,
isobutyl acetate and isoamyl acetate (Table I). The
studies focussed on the production of ethyl butyrate,
useful in pineapple-banana flavors, owing to its large
market demand of over 140,000 kg annually (<u>28</u>) and
selling price for the natural ester ranging upwards of

$150/kg. The system was found to produce significant amounts of this ester and as well had an excellent operational stability (upwards of a month operation).

Table I. Ester production by immobilized C. cylindracea lipase using a range of substrates. The molar concentration is that of ester in the heptane phase after 24 h. Percent molar conversion is that of initial acid to ester after 24 h

ESTER	(M)	CONVERSION
Ethyl propionate	0.19	76
Ethyl butyrate	0.25	100
Ethyl hexanoate	0.09	44
Ethyl heptanoate	0.21	84
Ethyl octanoate	0.26	100
Ethyl laurate	0.13	52
Ethyl isobutyrte	0.18	72
Ethyl isovalerate	0.01	3
Isobutyl acetate	0.06	25
Isoamyl acetate	0.06	24
Isoamyl butyrate	0.22	91

SOURCE: Reprinted with permission from ref. 27. Copyright 1987 Science and Technology Letters.

The key finding was the requirement for intermittent hydration of the immobilized enzyme for long term operational stability (Figure 4). Without this hydration protocol, upwards of 90% of the initial activity was lost after only 2 cycles (ca. 48 h).
    Another even more recent advance for flavor processing involves the use of supercritical fluids. Supercritical fluids, including supercritical CO_2, created under elevated pressures, exist in a nebulous region between that of a liquid and that of a gas. The ability to continuously adjust the solubility of the desired component to be extracted by modification of pressure and/or temperature has led to this technology being quickly adopted by food and flavor/fragrance industries (decaffeination, oil extraction, etc.). As well, the concern over the toxicity of certain solvents, has encouraged the search for an alternative extractive process. Supercritical CO_2 has been the 'solvent' of choice by most flavor industries employing this technology since it is inert, nonflammable, nonexplosive and leaves no residue in the final product. Moreover CO_2 is readily available in large quantities and high purity and of great importance it is probably the next cheapest solvent to water (under 10 cents/kg). Additionally, as the critical temperature of CO_2 is only

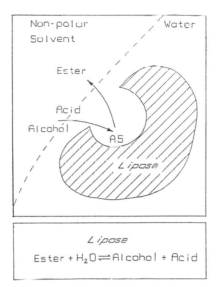

Fig. 3. Production of esters in a non-aqueous lipase system. Water present in low levels around enzyme. Reversal of equilibrium occurs to favor formation of esters. Active site = AS.

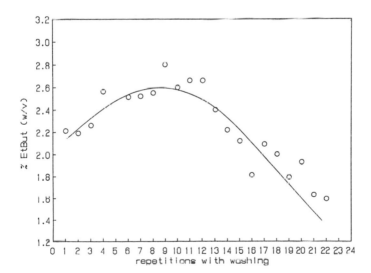

Fig. 4. Repeated use of immobilized lipase in non-aqueous system. Between each repetition the immobilized enzyme adsorbed to a silica gel support matrix was washed with water. Fresh substrate (ethanol, butyric acid) was added in hexane. One cycle represents ca. 24 h.

31°C at 74 bar pressure, thermally labile flavor
components such as jasmine can be readily extracted from
the botanical source. For the most part, it is these
attributes described above that the flavor industry has
found attractive for flavor compound extraction that
hasled biotechnologists to adapt this technology to
biocatalysis. Pioneering studies involving the use of
lipase in supercritical CO_2 have demonstrated, as
detailed above for solvent-based non-aqueous systems,
that hydrolytic reactions can be suppressed and
synthesis can be significantly enhanced (28, 29). One
group (29) found that the reaction rate in this type of
system was upwards of 270 times that (measured in more
product/mole enzyme.second) compared with an aqueous
system. Undoubtedly, more novel biotechnological
process routes will become available and as illustrated
by these examples, even once considered 'extreme'
environments could facilitate natural flavor
production.

REGULATORY CONSIDERATIONS-AN UPDATE

In the U.S., FEMA (Flavor and Extract Manufacturers'
Association) is in the process of drafting new
guidelines for natural flavor/aroma chemicals and
accepted processing methods (12). Apart from the
accepted use of fermentation, heat reactions and
enzymes, it is expected that inorganic acids and bases
will be allowable for use as catalysts. Obviously this
could allow for production of certain natural flavors,
in the absence of enzymes, under appropriate conditions,
however the requirement for natural substrates will
remain.
       Although regulatory agencies, such as the FDA, are
trying to keep pace with this rapidly advancing
technology many potential bottlenecks and 'grey' areas
persist primarily in interpretation of the regulatory
requirements. The FDA states that its policies are
sufficiently comprehensive to apply to those involving
biotechnology. Although, statements (30) such as
"...the use of new microorganisms found in a food such
as yogurt could be considered a food additive. Such
situations will be evaluated case-by-case..." could have
significant ramifications. For example, if a new
microorganism, possibly manipulated using rDNA
technology, is added to a food for flavor improvement, a
food additive petition (FAP) could be necessary. This
would add much to product development costs and possibly
delay effective early market penetration. Normally a
GRAS (Generally-Recognized-As-Safe) status for a food
substrate allows for immediate use in food products
without regulatory clearance. A major concern is that
the use of rDNA techniques could cause a substrate, that
was normally classified as GRAS, to be seen as a food

additive thereby necessitating regulatory clearance before use (31). The FDA has indicated that the use of an animal food substance prepared by rDNA technology must be approved by an FAP (32). Although, if the substance produced through this new technology can be shown to be identical to a GRAS substance, it probably would be cleared for use without being labelled as a food additive. Normally an FDA GRAS status is identified with the substance itself and its intended limitations on use, without mention as to its method of manufacture. It appears that the FDA will, however, require clearance by either a food additive regulation or by a regulation affirming GRAS status for products derived through rDNA techniques (33).

A number of the compounds used for flavoring materials are quite volatile, such as acetaldehyde, and are readily recoverable in pure form from existing microbial cultures or their derived enzyme systems. Thus the use of non-GRAS microbial or enzyme systems, within reason, could presumably be used to produce volatile GRAS flavor substances. Although it should be noted that the method of manufacture could be taken into consideration if any question of purity of a GRAS substance produced by a non-GRAS biological system is utilized.

In the systems developed by us described above, attention has been directed at compliance with regulatory considerations, primarily FDA. The production of ethyl acetate and acetaldehyde, both GRAS flavor substances, was accomplished using a GRAS yeast, C. utilis. The medium containing basic salts and either glucose or fermentation ethanol should not raise any questions regarding potential toxicities. The end products being very volatile can be readily isolated even if there had been any concern over the microbial system being employed. Additionally, the spent yeast produced in this process is one of only three GRAS yeasts allowed for use in food products destined for human consumption, and can therefore give a by-product credit to this process as discussed earlier. It is tempting to go into nature to isolate new microbial systems but the potential costs can be staggering from a regulatory standpoint. A good example of this is the $60 million that Rank Hovis McDougall's company in the U.K. spent developing and seeking regulatory approval for a Fusarium SCP process (34).

The non-aqueous lipase system for flavor esters developed by our group used components and preparative techniques for enzyme immobilization, that would not only be cost effective and simple but also meet regulatory requirements. The enzyme could have been immobilized by a number of methods however for the intended application only (i) adsorption (ii) ionic bonding or (iii) glutaraldehyde cross-linking would be

appropriate. Adsorption of the lipase to silica gel was
chosen as it is an extremely simple method eliminating
the need for potentially dangerous chemicals. Strict
regulations exist for food-related processes especially
where solvents are involved. Hexane was chosen for
thisprocess as it is used almost universally for plant
material extraction protocols (35). A typical allowable
maximum solvent residue in natural extractives is 25 ppm
(36). Finally, the lipase from C. cylindracae had also
had a precedent set for use in food (cheese ripening and
butter fat modification).
        Industries dealing with food or pharmaceutical
products are beginning to incorporate biotechnology into
their product development schemes although the routes
being taken are rather limited. Large deviation from
the use of a select short list of GRAS microorganisms or
their derived enzymes is not common practice (8). Where
possible 'classic' genetic and/or physiological
manipulation of these organisms is done in order to
facilitate regulatory compliance and approval. As
industry and government regulatory agencies work closer
together to clear up the concerns over the use of 'new'
biotechnology, including genetic engineering, more
applications will emerge. Apart from compliance with
government regulations, it is critical that the industry
itself set its own high standards as their liabilities
are not necessarily protected by government guidelines
(12).

INDUSTRIAL INVOLVEMENT IN BIOTECHNOLOGICAL FLAVORS DEVELOPMENT

Unlike the pharmaceutical industry, food processing
industries tend to be market driven rather that
technology driven (7) and is much more receptive to
trends such as the demand for natural flavors. It is
anticipated that there will be more investment in
biotechnological flavor development especially as the
technology proves itself and that existing technology
needs can be satisfied by these new approaches.
        The majority of large companies' involvement in
biotechnology is in the form of R&D contracts to small,
entrepreneurial companies. This is a strategic approach
which allows for a minimal investment of capital in
order to test if the technology developed can be
applicable to their needs (36). If the impact is
significant, an in-house program would then evolve. A
number of companies have taken this approach including:
Firminech with DNA Plant Technology for improved
production of flavors; W.R. Grace with Synergen for
development of microbial systems for flavors; General
Foods, although they do have an in-house program, have
established links with ESCAGENETICS. F&C International
(Canada) Ltd. has established a research collaboration

with the National Research Council, a Canadian Federal Government research organization.

With the projected increase in natural flavors demand expected to grow at an annual rate of about 20% (12) and the current average price of about $150-200/kg more companies are expected to join in this endeavor. At present only about half a dozen companies are involved in developing natural flavor/aroma chemicals with some retaining a number of their products for captive use in their own flavor blends. It has been estimated (12) that one of the larger companies has upwards of 25-30 natural compounds developed.

FUTURE DIRECTIONS

A number of areas will see significant increases in research activity in the future and include:

● A greater understanding of the capabilities of biocatalysts will allow for an upgrading and increased commercial exploitation of many unutilized biological substances such as certain components of essential oils.

● The trend toward more healthy lifestyles has also encouraged a demand for unsaturated and low fat diets which has increased the use of plant rather than animal products. The preference of 'meaty' flavors will necessitate more use of fermentation-derived flavoring agents such as 5'-nucleotides including inosinate and guanylate, along with hydrolyzed yeast. Overproduction of enzymes to do the hydrolysis, by rDNA techniques, could allow for use of yeast extracts or autolysates directly without the need for purification of RNA by chemical means (20). Recently, a group in Scotland has succeeded in cloning genes for this purpose in yeast which can convert RNA of any origin to these flavor-enhancing compounds (37).

● Potential development of 'industrially robust' enzyme-mimetic systems to allow for specific biotransformations. These mimics would be catalytically active but not be dependent upon the amino acid backbone of existing enzymes.

● An increased use of plant cell culture, for production of plant-unique flavors, will appear as more is understood on underlying control and synthetic mechanisms. Expression of certain plant genes in microorganisms will also be done. An early example of this is the 'taste active' plant protein thaumatin (2500-3000 times sweeter than sucrose) being expressed in microbial systems. For industrial scale production many problems exist with large scale plant cell

cultivation as a result of sensitivity to shear forces
(cell breakage), and their tendancy to clump thereby
limiting mass transfer (nutrient limitation). Enabling
technologies such as appropriate bioreactor design and
suitable production media must be developed (10).

● Numerous food products will be upgraded to
'value-added' products by the incorporation of
flavorings with appropriate functional properties
extending beyond flavoring. An important area of
research will be in those flavors imparting antioxidant
and preservative function (38). Sage and rosemary
extracts have been shown to function as antioxidants
while those of nutmeg, mace and bay leaf impart
preservative effects inhibiting growth of Clostridium
botulinum. Microbial or enzyme production of specific
flavors with these properties should be possible.

● Certain volatile aroma compounds have been shown to
bring about positive changes in mood or general
well-being and is encompassed in a new discipline called
aroma therapy. It is unlikely that large demands on
certain aroma-therapeutic compounds could be satisfied
by traditional technologies. Biotechnological
approaches could satisfy these requirements for aroma
therapeutics.

● The future for production of natural flavors through
the use of biotechnological means holds much promise.
Researchers and industries interested in this exciting
area however, will only be successful if they maintain
an awareness of market trends, and of equal importance,
knowledge of regulatory requirements.

Literature Cited

1.   Newell, N. Gen. Eng. News 1986, 6, 55.
2.   Hacking, A.J. Economic Aspects of Biotechnology;
     Cambridge Press, N.Y., 1986, p 29.
3.   Godfrey, T. In Industrial Enzymology; Godfrey, T.;
     Reichelt, J, Eds., Nature Press, N.Y., 1983; pp
     305-314.
4.   Rijkens, F.; Boelens, H. Proc. Int. Symp. Aroma
     Res., 1975, p 203.
5.   Maarse, H. In Volatile Compounds in Food -
     Quantitative Data; TNO, Netherlands, 1984, Vol.
     1-3.
6.   Bedoukian, P.Z. Perfum. Flavor. 1985, 10, 1-27.
7.   Horton, H.W. Food Technol. 1987, 41, 80.
8.   Armstrong, D.W., Yamazaki, H. Trends Biotech. 1986,
     4, 264-268.
9.   Newell, N.; Gordon, S. In Biotechnology in Food
     Processing; Harlander, S.K.; Labuza, T.P., Eds.;
     Noyes Publ., Park Ridge, NJ, 1986, pp 297-311.

10. Whitaker, R.J.; Evans, D.A. Food Technol. 1987, 41, 86.
11. Tahara, S.; Fujinara, K.; Mizutani, J. Agr. Biol. Chem. 1973, 37, 2855.
12. Anonymous. Chem. Marketing Reporter (5 Oct.), 1987, p 5.
13. Tanabe Seiyaky Co. Ltd. 1979, Jap. examined patent 5, 407, 859.
14. Armstrong, D.W. In Biogeneration of Aroma Compounds; Parliment, T., Croteau, R., Eds.; American Chemical Society, Washington, DC, 1986, pp 254-265.
15. Armstrong, D.W.; Martin, S.M.; Yamazaki, H. Biotech. Lett. 1984, 6, 183-188.
16. Byrne, B.; Sherman, G. Food Technol. 1984, 57.
17. Dziezak, J.D. Food Technol. 1987, 104.
18. Dhavlikar, R.S.; Albroscheit, G. Dragoco Rep. 1973, 12, 251-258.
19. Anonymous. Chem. Marketing Reporter (14 March), 1988, 38.
20. Trivedi, N.B. Gen. Eng. News 1986, 6(2), 1.
21. Arbige, M.V.; Freund, P.R.; Silver, S.C.; Zelko, J.T. Food Technol. 1986, 40, 91.
22. Godfrey, T. In Industrial Enzymology: The Application of Enzymes in Industry; Godfrey, T.; Riechelt, J., Eds.; Nature Press, N.Y., 1983, pp 424-427.
23. Nelson, N.; Racker, E. Biochemistry 1973, 12, 563.
24. Seo, C.W.; Yamada, Y.; Okada, H. Agric. Biol. Chem. 1982, 46, 405-409.
25. Gillies, B.; Yamazaki, H.; Armstrong, D.W. In Biocatalysis in Organic Media; Laane, C.; Tramper, J.; Lilly, M.D., Eds.; Elsevier, Amsterdam, 1987, pp 227-232.
26. Laane, C.; Boeren, S.; Hilhorst, R.; Veeger, C. In Biocatalysis in Organic Media; Laane, C.; Tramper, J.; Lilly, M.D., Eds.; Elsevier, Amsterdam, 1987, pp 65-95.
27. Gillies, B.; Yamazaki, H.; Armstrong, D.W. Biotechnol. Lett. 1987, 9, 709-714.
28. Anonymous. Bioprocess. Technol. 1987, 9(4), 1.
29. Anonymous. Bioprocess. Technol. 1987, 9(9), 6.
30. Title 49 Federal Reg. 50878 (31 Dec.), 1984.
31. Korwek, E.L. Food Technol. 1986, 40, 67.
32. Title 49 Federal Reg. 50879 (31 Dec.), 1984.
33. McNamara, S. In Biotechnology in Food Processing; Harlander, S.K.; Labuza, T.P., Eds.; Noyes, Park Ridge, NJ, 1986, pp 15-27.
34. Hacking, A.J. Economic Aspects of Biotechnology; Cambridge Press, U.K., 1986, p 29.
35. Bosman, R.J. In Safe Uses of Solvents; Collins, A.J.; Gluxon, Eds.; Academic Press, N.Y., 1982, pp 25-34.

36.   Newell, N. Gen. Eng. News 1986, 6(2), 5.
37.   Anonymous. Biobulletin 1987, 5(1), 2.
38.   Williams, S.K., Brown, W.L. Food Technol. 1987,
      41(6), 76.

RECEIVED November 8, 1988

PERCEPTION

Chapter 10

Neurophysiology and Stimulus Chemistry of Mammalian Taste Systems

James C. Boudreau

Health and Science Center at Houston, Sensory Sciences Center, University of Texas, 6420 Lamar Fleming Avenue, Houston, TX 77030

Single unit recordings were taken from sensory ganglion cells innervating oral taste buds in the cat, dog, rat, and goat. Neurons were divided into 9 groups largely according to stimulus chemistry. A sodium-lithium system was seen in the rat and goat but not the cat and dog. Amino acid responsive neurons were seen in all species except the goat, with major species differences. Amino acid responsive neurons were also, except for the cat, responsive to sugar. A nucleotide system was seen only in the cat. Acid (Brønsted) responsive neurons were seen in all species, but the cat and dog acid taste systems were different from others. A system responsive to furaneol and other compounds present in fruit was seen only in the dog. A system exclusively responsive to alkaloids was found in rat and goat. Type of taste systems present can to a certain extent be related to species' ecology and dentition.

Flavor chemists typically subdivide the perception of food into three types of sensations: taste, smell and flavor. This latter category almost invariably consists of sensations during consumption. The flavor sensations are considered largely to arise from the stimulation of smell receptors, although research has not demonstrated this to be so. From a biological and physiological point of view, these flavor sensations have little reality. In biology, food odors have been found to have little to do with consumption, being primarily concerned with

orientation. Consumption is under the control of
contact chemoreceptors or taste. Flavor sensations must
then consist primarily of taste sensations and/or
sensations arising from the simultaneous stimulation of
both taste and smell receptors.

In this report the neurophysiology of mammalian
taste systems is reviewed with especial attention to
stimulus chemistry. The neurophysiology described is
primarily that from our laboratory, since we have been
among the few neurophysiologists concerned with stimulus
chemistry. The animals that have been investigated in
detail are the cat, dog, goat and rat. Work on other
animals is included where comparisons are viable.

Anatomy and Physiology

Four cranial nerves subserve the sense of taste, three
of these (facial, glossopharyngeal and vagus) innervate
taste bud systems (Fig. 1) and one (trigeminal) supplies
free nerve ending receptors. Both of these types of
receptors respond to chemical stimuli. Only the taste
bud systems of the facial and glossopharyngeal nerves
have been studied in sufficient detail with many food
compounds.

The neurophysiological preparation used was metal
electrode recordings from the sensory ganglion cell
bodies in the geniculate (facial nerve) and petrosal
(glossopharyngeal) ganglia of anesthetized animals (Fig
1). This preparation permits long term extracellular
recordings from sensory neurons with their peripheral
and central extremities intact. Neurophysiological
measures taken included spontaneous and evoked activity,
and receptive field papillae system mapping with latency
measures.

The spike trains recorded from first order taste
neurons have some unusual characteristics (Fig. 2). All
taste neurons have a certain level of spontaneous
activity. This spontaneous activity is often of a highly
complex nature. "Bursting", in which the spikes appear
in short relatively fixed intervals are common, and
"grouping" in which a pseudo-discharge appears is also
not unusual. When excited by optimal stimuli two types
of discharge may occur. In one, the spikes are
tonically occurring with usually a fairly rapid decline
in the first few seconds. In the other type, the spikes
may appear in groups, often after a long latency. The
first type of discharge is common to most geniculate
neurons; the second to some geniculate ganglion units
and most petrosal ganglion units. Examples of evoked
discharges recorded from peripheral sensory ganglion
cells are presented in Figure 2.

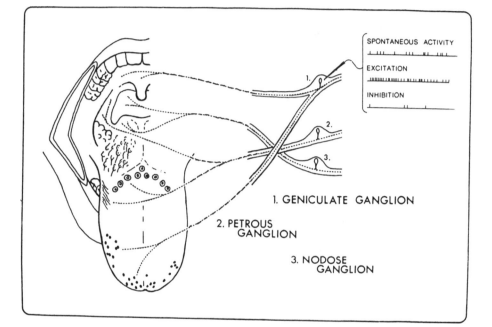

Figure 1. Diagram of the three cranial nerves and
associated sensory ganglia that innervate taste buds.
As illustrated, electrical recordings were taken from
single neurons in the ganglia. Geniculate ganglion
in facial nerve; petrosal in glossopharyngeal; nodose
in vagus.

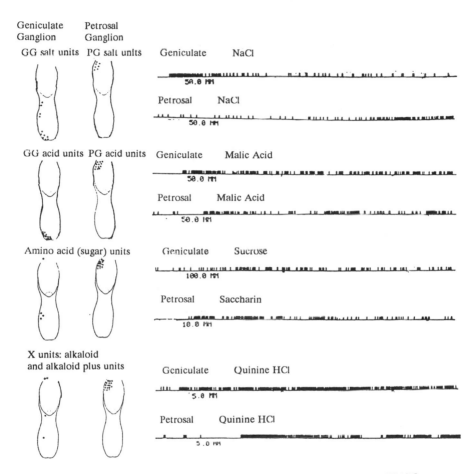

Figure 2. Taste systems of the rat geniculate (GG) and petrosal (PG) ganglia. Location of receptive fields indicated by a dot on tongue for each neuron studied. Examples are shown of elicited spike discharge for neurons from the six different neural groups identified.

Neural Groups

In every animal studied, the neurons could be divided
into a number of neural groups according to their neurophysiological characteristics and the chemical stimuli
to which they were responsive. The neural groups that
been described in the mammalian geniculate ganglion (GG)
and petrosal ganglion (PG) are listed in Table 1, along
with some of their characteristics. Geniculate ganglion
neurons have been studied in four species (1), but
petrosal ganglion units have been studied only in the
rat (2). The geniculate ganglion units can be placed
into at least seven different neural categories, but a
group may be absent from one species or may respond to a
somewhat different stimulus array. The neurons in the
rat petrosal ganglion have been tentatively divided into
four distinct groups, but two of these groups are
similar to rat geniculate ganglion groups. All told, at
least nine distinct peripheral taste systems can be
distinguished in the four species studied. Most of
these neural groups have also been distinguished in
peripheral fiber recordings in other laboratories (1).
    The main criteria used to classify the units in
Table I were stimulus response measures; i.e., the units
discharged or were inhibited by different chemical compounds. In addition, other criteria were used to
supplement the chemical stimulus response
differentiation. Thus, the two main groups in the cat
(acid units and amino acid units) can also be
differentiated by spontaneous activity measures, latency
to electrical stimulation, area of tongue innervated,
and differential response to solution temperature (3-5).
This comparative work has led to a modular view of
peripheral taste systems in which the different neural
groups are seen to have distinct receptors responding to
distinct types of chemical signals (e.g., Brønsted acids
and Brønsted bases), with either excitation or
inhibition. The stimulus chemistry of these groups will
be briefly described.

Salt Responsive Units. One of the neural groups with
the simplest stimulus chemistry is the GG salt system
found only in the geniculate ganglion of the rat and
goat. These units are only responsive to sodium or
lithium salts. When a series of Cl salts with different
cations are examined, only those with Na and Li elicit
large responses (Fig. 3). Na and Li are effective with
other anions as well, although responses are largest
with I and F (6).
    The only other group of neurons responsive
exclusively to salts was the rat PG salt unit group
(Fig. 3). These units of the petrosal ganglion
responded to a variety of Cl salts, not showing the Na,

Table I
Mammalian Peripheral Neural Taste Groups

Geniculate Ganglion (Facial Nerve): GG. Petrosal Ganglion
(Glossopharyngeal Nerve): PG

Group	Species	Stimuli
1. GG Salt System	Rat and Goat only	Na^+ and Li^+
2. GG Acid System	Rat and Goat different from Cat and Dog	Brønsted acids
3. GG Amino Acid System	Cat and Dog	Proline, Cysteine, Hydroxyproline, Lysine, Alanine
4. GG Nucleotide System	Cat only	ITP, ATP, etc.
5. GG Furaneol System (Probably mainly PG)	Dog only	Furaneol, Ethyl Maltol Methyl Maltol
6. PG Amino Acid System (also in GG)	Rat	Sugar, Saccharin, Amino Acids
7. PG Alkaloid System (also in GG)	Rat and Goat	Atropine
8. PC Acid System	Rat	Restricted set of carboxylic acids
9. PG Salt System	Rat	KCl, $CaCl_2$, $MgCl_2$, NaCl

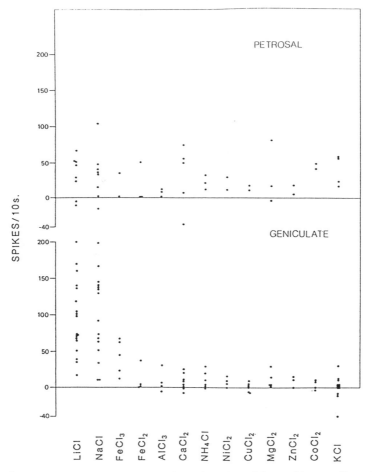

Figure 3.   Responses of rat GG salt units and rat PG salt units to a series of chloride salts.   Each point represents the spike response in a 10 second period to a 50mM solution.   Note the exclusivity of response of GG salt units to NaCl and LiCl.

Li exclusivity shown by GG salt units. In addition, they exhibited low discharge rates and sluggish evoked discharge patterns.

Nucleotide Responsive Units. Certain of the cat units were observed to discharge only to nucleotides and other phosphate compounds. In addition to these nucleotide units, some other cat units also discharged to nucleotides and other substances. In general, the di- and tri-phosphate nucleotides were the most stimulating for the nucleotide units although both AMP and IMP elicited respectable responses. Tetrasodium pyrophosphate was a strong stimulus and sodium phosphate a moderate one. These units exhibited long latencies to electrical stimulation, low spontaneous activity rates, and "grouping" evoked discharge patterns. No specific regional distribution of receptive fields was observed.

Furaneol Responsive Units. Found only in the dog were a small number of units responsive to a variety of plant compounds known to be intensely sweet for the human. Especially active was the compound furaneol [2,5 dimethyl-4-hydroxy-3(2H) furanone] and the closely related ethyl and methyl maltol. Also stimulating were ammonium glycyrrhizinate and (slightly) neohesperidin dihydrochalcone. Some units were also responsive to quinine. No response was shown to either amino acids or sugars, nor were salts or acids stimulating. These furaneol units were the only units in any species responsive to intense sweeteners. Furaneol and other compounds were tested on many cat units and some rat units but no discharge was evoked. The dog units possessed small fibers and displayed "grouping" discharges, often with long latencies. It is quite likely that these units are representative of a larger population of neurons in the petrosal ganglion of the glossopharyngeal nerve, a preparation not studied in the dog.

Acid Responsive Units. All species possessed an acid taste system although this system was not identical from species to species. The system was labeled "acid" because the most stimulating compounds were Brønsted acids and the least stimulating were Brønsted bases. The most excitatory compounds were carboxylic acids for all species. Also stimulating, but at a variable rate, were phosphoric acids and a small number of nitrogen compounds functioning as Brønsted acids. Histidine, functioning as a Brønsted acid, was active in all species. The compounds with phosphoric acid groups were least active on the rat and goat. Salts such as NaCl and KCl were active on the rat and goat though less so. The acid units in the cat were studied in the most

detail. It was found that imidazole was even more stimulating than histidine. A small group of nitrogen
heterocycles when protonated, were the most excitatory
compounds for the cat. The heterocycles, imidazole,
thiazolidine, and pyridine with their relatively high
pK's were extremely exciting at a pH of 7.0. In the
cat's normal diet of meat, a pH below 5.5 is rarely
encountered, rendering most carboxylic and phosphoric
acids nonstimulating. Present in large quantity in
animal tissues in free form however are histidine dipeptides: anserine, carnosine and ophidine, depending on
animal species. Dog acid units were almost identical to
those in the cat.

Present in the rat petrosal ganglion was another
set of acid units responsive primarily to certain
carboxylic acids. Unlike the cat (7), the rat was unre-
sponsive to some carboxylic acids even though they were
in low pH solutions (2). Possibly the same is true for
the rat GG acid units which were not investigated in as
much detail. PG acid units, unlike all other acid unit
groups, responded in a "grouping" discharge fashion.
Goat acid units seemed in between carnivore and rat acid
units, being more responsive to phosphate compounds.

Alkaloid Responsive Units. Present in the rat and in
the goat were units which were responsive primarily to a
small group of alkaloids. These units were found in the
geniculate ganglion where they were few and innervated
the back part of the tongue. They were found in larger
number in the rat petrosal ganglion. These units
exhibited long latencies to electrical stimulation,
indicating small fiber diameters and displayed "grouping"
evoked discharge patterns. The rat alkaloid units were
maximally discharged to atropine, quinine, colchicine
and sparteine. The goat units were maximally discharged
by pilocarpine, quinine and colchicine. Few other non-
alkaloids were active although $CaCl_2$ was stimulatory for
some rat and goat units. A few units in the cat were
maximally discharged by alkaloids (mainly quinine and
brucine) but they were not studied with an array of
alkaloids.

Amino Acid Responsive Units. Found in the geniculate
ganglion of the cat, dog, and rat, but not in the goat,
are neural groups highly responsive to amino acids (Fig.
4). The amino acid units of the dog and rat, but not of
the cat, are also responsive to sugars. The amino acid
units of all three species are also responsive to
nucleotides but less so in the rat. The rat amino acid
units are distinct from those in the carnivore in that
different amino acids are maximally stimulatory, and the
discharge rates are usually much lower. An amino acid
group of neurons was also detected in the rat petrosal

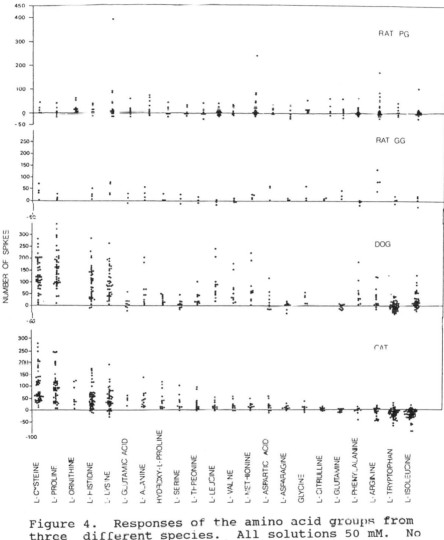

Figure 4. Responses of the amino acid groups from three different species. All solutions 50 mM. No amino acid units were seen in the goat.

ganglion where, unlike in the geniculate ganglion, these
units formed the most populous group. The rat
geniculate ganglion amino acid units are similar in
general in their stimulus response properties to the rat
petrosal ganglion amino acid units.

Cat amino acid units are essentially discharged by
two distinct types of compounds: those containing
phosphate groups, such as sodium phosphate, tetrasodium
pyrophosphate, and all di- and triphosphate nucleotides;
and certain amino acids. Monophosphate nucleotides
elicited little response from amino acid units. The
most effective amino acids in eliciting discharge were
L-proline, L-cysteine, L-ornithine, L-lysine, L-
histidine, and L-alanine. Certain amino acids such as L-
tryptophan, L-isoleucine, L-leucine, L-arginine, and L-
phenylalanine tended to inhibit cat amino acid units.
The inhibitory property of the L-amino acids has been
related to the hydrophobicity of their side chains

Cat amino acid units have been studied with a
variety of stimulus solutions including natural foods
such as chicken and liver (8, 9). The most excitatory
compounds, as indicated above, tended to be either compounds with a phosphate group or compounds with a
nitrogen group. The D amino acids tend to be less
stimulatory than the L forms. The most effective
stimuli found included small heterocyclic nitrogen compounds such as pyrrolidine. Inhibitory compounds were
mostly alkaloids, nucleotide bases and certain
heterocyclic nitrogen compounds. The response of cat
amino acid units to nitrogen heterocycles could be
related to two structural and chemical factors: (i) a
steric factor (in particular ring size) and (ii) the
relative basicity of the heterocycles as indicated by
pKa values. Cat amino acid units were also discharged
by NaCl and KCl solutions, but the thresholds were above
50 mM.

Although the most stimulatory amino acids were
identical in the dog (e.g.,L-cysteine, L-proline, L-
lysine, L-histidine and L-alanine), interspecies
differences could be related to the side chain
properties of the amino acids. Thus, amino acids with
hydrophobic side chains were normally inactive or inhibitory in the cat, but were often excitatory in the dog.
Conversely, amino acids with acidic side chains tended
to be somewhat more excitatory in the cat.

The response of the rat amino acid units to amino
acids was quite distinct from that of the cat and dog.
Little response, for instance, was elicited from rat
amino acid units by most of the di- and triphosphate
nucleotides, and sodium phosphate was inhibitory. The
most effective amino acid for rat units was L-arginine,
a compound inhibitory in the cat and a minor stimulus in
the dog, followed by L-lysine and L-aspartic acid. L-

proline was largely inactive in the rat. Few of the rat
amino acid units discharged at high rates. Rat units
were also responsive to sugars and saccharin.

Summary of Mammalian Neural Groups

The different neural groups distinguished in the
geniculate and petrosal ganglia are summarized with
respect to species in Table II. The animals studied in
the geniculate ganglion have been supplemented with
three species studied only in the chorda tympani: the
hamster (10, 11), the squirrel monkey (12) and the
macaque (13). The amino acid units in the two primates
seem to represent the two different types of amino acid
units seen in the ganglion preparation. The squirrel
monkey amino acid units seem quite similar to dog amino
acid units even though the investigators themselves
classify them as salt units. The macaque units on the
other hand display the unusual grouping discharge
patterns shown by rat amino acid units.   The human
is included in this table because the different human
sensations seem to represent psychophysical signs of
excitation or inhibition of different neural groups
(14). On the basis of chemicals active, the human acid
units seem more like those of the cat and dog than the
rat or goat (7, 14). The human sodium system seems
identical to that in the rat, hamster and goat (15).
The human clearly possesses a facial nerve amino acid
system similar to the carnivore (16) and a petrosal
system similar to the dog furaneol system (14, 17). The
human also possesses a glutamate system, yet undetected
in any experimental mammal (18, 19).

Discussion

The modular taste systems summarized for mammals in
Table II are quite similar to the modular taste systems
that have been observed for invertebrates, such as
lobsters and crayfish (20, 21). The most extensive
invertebrate taste research has been performed on
caterpillars (22, 23). In 20 different species of
caterpillars, 12 different neural groups were
distinguished.

Viewed in terms of neural groups, the experimental
animals detailed here constitute a diverse group of
organisms. The rat and the hamster seem to possess
identical geniculate ganglion systems. Should the rat
and hamster prove to be representative of rodents in
general, this sodium, acid, amino acid-sugar taste
system may be common to most or all rodents (of which
there are around 2400 species). The rodent taste system
is also quite similar to that of the goat; although no
amino acid-sugar system has yet been detected in the

Table II
Summary of Mammalian Peripheral Neural Taste Groups
(See Text)

Neural Groups	Species							
	Cat	Dog	Rat	Goat	Hamster	Sq. Monk	Mac. Monk	Man
Facial								
GG Salt (Sodium)			X	X	X		X	X Salty
Amino Acid Cat type	X	X				X		X Sweet1
Acid, Cat Type	X	X						X Sour
Acid, Rat Type			X	X	X			
Nucleotide	X							
Glossoph.								
Amino Acid, Rat Type			X^a		X^a	X^a(?)		
Furaneol		X^a						X^a sweet2
PG acid			X					
PG salt			X					
Alkaloid	?		X^a	X^a				
Glutamate[b]								X^b umami

a: Also in facial
b: Psychophysics only

goat, the alkaloid system, sodium system and GG acid system of the goat are like those of the rat. Perhaps the taste systems of many mammals capable of living on plant foods contain basic similarities.

The cat and the dog, on the other hand, possess taste systems that have little in common with rodents and goats. Not only do they have no sodium system, but their acid and amino acid systems are also markedly distinct. Although the cat and the dog have two systems, the acid and amino acid systems, in common, both also possess a taste system which the other does not: the cat a nucleotide system and dog a furaneol system.

The primates have been inadequately studied, but those two with adequate single unit data suggest that the organization of primate taste systems is no simple matter. It is not obvious for instance, why the squirrel monkey may have an amino acid system like a carnivore and the macaque one like a rodent. The human taste system further complicates matters since man can best be viewed as a composite, having a sodium system like the rat and goat, carnivore acid and amino acid systems, a furaneol system like the dog and a glutamate system unlike any other mammal studied (14).

The compounds active on both vertebrate and invertebrate taste systems constitute a select group of low molecular weight compounds. The compounds include organic acids, salts, nucleotides, amino acids and a variety of secondary compounds, notably alkaloids but also others, including here furaneol and ethyl and methyl maltol. Just why certain of these compounds are active on taste systems is often a moot point. The significance of none of the acid systems, for instance, is obvious from an ecological standpoint, nor is it apparent why certain acids are so potent. It is also not clear why the two amino acid systems are so distinct, nor why proline and cysteine should assume such a large role in the carnivore taste system.

The taste systems which are ecologically obvious, however, are the GG sodium system and the dog furaneol system. The sodium system is not present in carnivores but is present in herbivores and omnivores. The importance of this system in the rat and goat cannot be overemphasized since half of the taste neurons in the geniculate ganglion are devoted to sodium sensing. The presence of a sodium system in animals that may subsist entirely on plant substances is quite obvious since Na is often present in minuscule quantities in most plants (24). Both the rat and goat exhibit a salt hunger and can with saline solutions regulate their sodium intake to supply their sodium need. Although the dog (and related canines) may subsist for fairly long periods of

time on fruit or other plant substances, it cannot
regulate its sodium intake by taste (25).
The dog units were labeled furaneol units because
this compound is found in large quantity in many fruits
(26). Besides being intensely sweet, this compound also
has a fragrant odor and is a character impact compound
for many fruits. It is believed that this dog furaneol
taste system is specific for fruit and is linked with
the seed dispersing function of the dog. The presence
of this taste system and its absence is readily
detectable in the natural eating behavior of canines and
felids. In a natural environment canines will supplement their small animal diet with fruit of the season,
unlike felids. Nucleotide responsive units are
relatively rare in taste systems. The only other vertebrate nucleotide taste system that has been described is
in the puffer fish (27). This fish facial nerve taste
system, like that in the cat, also responded to a wide
variety of nucleotides and to inorganic phosphate
compounds. In invertebrates, nucleotide taste systems
have been described for blood sucking animals where they
are common (28).

Literature Cited

1. Boudreau, J.C.; Shivakumar, S.; Do., L.T.; White,
    T.D.; Oravec, J.; Hoang, N.K. Chem. Senses 1985, 10,
    89-127.
2. Boudreau, J.C.; Do, L.T.; Shivakumar, L.; Oravac,
    J.; Rodriquez, C.H. Chem. Senses 1987, 12, 437-458.
3. Boudreau, J.C.; Alev., N. Brain Research 1973, 54,
    157-175.
4. Ishiko, N.; Sato, Y. Jap. J. Physiol. 1973, 23, 275-
    290.
5. Nagaki, J.; Yamashita, S.; Sato, M. Jap. J. Physiol.
    1964, 14, 67-89.
6. Boudreau, J.C.; Hoang, N.K.; Oravac, J.; Do., L.T.
    Chem. Senses 1983, 8, 131-150.
7. Boudreau, J.C.; Nelson, T.E. Chem. Senses and Flavor
    1977, 2, 353-337.
8. Boudreau, J.C.; Anderson, W.; Oravec, J. Chem.
    Senses and Flavor 1975, 1, 495-517.
9. Boudreau, J.C.; White, T.D. In Flavor Chemistry of
    Animal Foods; Bullard, R.W. Ed.; ACS Symposium
    Series No. 67; American Chemical Socity: Washington,
    DC, 1978; pgs. 102-128.
10. Frank, M.J. J. Gen. Physiol. 1973, 61, 588-618.
11. Fran, M.J.; Bieber, S.L.; Smith, D.V. J. Gen.
    Physiol., 1988, 91, 861-896.
12. Pfaffmann, C.; Frank, M.; Bartoshuk, L.M.; Snell,
    T.C. In Progress in Psychobiology and Physiological
    Psychology; Sprague, M., and Epstein, A.N., Eds.;
    Academic Press: New York, 1976; 6: pgs. 1-27.

13. Sato, M.; Ogawa, H.; Yamashita, S., J. Gen. Physiol. 1975. 66, 781-810.
14. Boudreau, J.C. J. Sensory Studies 1986, 1, 185-202.
15. Boudreau, J.C. Chem. Senses 1984, 9, 341-353.
16. Boudreau, J.C. In Flavor of Foods and Beverages; Charalambos, G. Inglett, G.,Eds.; Academic Press: New York, 1978; pgs. 232-246.
17. Dubois, G.E.; Crosby, G.A.; Lee, J.F.; Stephenson, R.A.; Wang, P.C. J. Ag. Food Chem. 1981, 29, 1269-1276.
18. Boudreau, J.C. In Umami, a Basic Taste; Kawamura, Y., Kare, M.R. Eds.; M. Decker, Inc.: New York, 1987; pgs. 201-217.
19. Yamaguchi, S. In Food Taste Chemistry; Boudreau, J.C., Ed. American Chemical Society: Washington, DC, 1979; pgs. 33-51.
20. Bauer, U.; Dudel, J. J.Comp. Physiol. 1981, 144 67-74.
21. Johnson, B.R.; Violet, R.; Borroni, R.; Atema, J. J. Comp. Physiol. A. 1984, 155, 593-604.
22. Schoonhoven, L.M. In Semiochemicals: Their Role in Pest Control; Nordlund, D.A., Ed.; J. Wiley and Sons; New York, 1981; pgs. 31-50.
23. Schoonhoven, L.M. In Perspectives in Chemoreception and Behavior: Proceedings in Life Sciences; Chapman, R.F., Bernays, E.A., Stuffolano, J.G., Eds.; Springer-Verlag: Berlin, 1987; pgs. 69-97.
24. Denton, D. The Hunger for Salt; Springer-Verlag: Berlin. 1982.
25. Fregly, M.J., In Biological and Behavioral Aspects of Salt Intake, Kare, M.R., Fregly, M.J., Bernard, R.A., Eds.; Academic Press: New York, 1980; pgs. 55-68.
26. Pickenhagen, W.; Velluz, A.; Passerat, J.P.; Ohloff, G. J. Sci. Food Agri. 1981, 32, 1132.
27. Hidaka, I.; Kiyohara, S.; Oda, S. Bull. Jpn. Soc. Sci. Fish. 1977, 43, 423-428.
28. Friend, W.G.; Smith, J.J.B. Ann. Rev. Enthomol. 1977, 22, 309-332.

RECEIVED September 23, 1988

Chapter 11

Temporal Aspects of Flavoring

P. Overbosch and W. J. Soeting

Unilever Research Laboratorium Vlaardingen, P.O. Box 114, 3130 AC Vlaardingen, Netherlands

The aroma of a food product is often measured by a sensory technique called descriptive profiling, in which flavour experiences are described by a panelist as a set of component impressions or sensations of varying degrees (1). Profiles do not hold explicit information about the temporal characteristics of a flavour, its persistence and the times of appearance of the individual notes. Nevertheless, the rank order of the attributes sometimes reflects, to some extent, the order of appearance of the corresponding impressions. Some notes are then said to be released "early", others "late".

However, there is little published experimental evidence that demonstrates a relationship between the temporal features of aroma perception and the stimulus concentration near the sensory receptors. In the following we describe some experiments that examine the issue directly and some theoretical ideas that appear to explain the results.

Psychophysical Measurement

When carrying out psychophysical measurements on a flavoured food we usually define the system as:

flavour/matrix   > sensory response

where the flavour/matrix is e.g. diacetyl in margarine and the sensory response is a magnitude score representing perceived intensity.

This view, however, is too simple. In reality we have to consider a stimulus-response system where the stimulus is defined as a concentration, not in the product but at the receptor sites and not as a single value but as a function of time. Likewise, the response should be measured as a function of time.

0097–6156/89/0388–0138$06.00/0

The system can then be formulated as follows:

flavour/matrix > stimulus (t) [psychophysical function (t)] > response (t)

To be able to understand this system we have developed:

- a methodology to measure the concentration of flavour as released from a matrix, at the nose, breath-by-breath;
- a time-dependent form of the psychophysical function relating stimulus to perceived intensity;
- an improved methodology to measure perceived intensity as a function of time (I/t).

Mass-spectrometric breath-by-breath analysis
_____

Breath-by-breath analysis of gases and volatiles is well known in medicine (1). The experimental techniques used, however, were not very well suited to our needs. For our purpose we needed a simple, reliable inlet system without extensive filtering and pressure reduction, but with a high sensitivity and short response times.

Therefore a (semi-) continuous measuring methodology, like MS was considered. Trace analysis by MS via a membrane separator was known (2), but the decay times of the signal precluded breath-by-breath analysis.

When the construction of the separator was studied more closely, it apppeared that the device traded response time for sensitivity. A fast response requires a small internal volume, but a high sensitivity requires a large membrane surface. Reducing the surface area and the internal volume resulted in very short response times and sufficient sensitivity (see Fig. 1).

The setup responsible for these improvements is depicted in Fig. 2.

Via two small glass pipes, one in each nostril, a small pump sucks 550 ml/min of air from the nose and past the membrane. The MS takes 20 data points/s and the result is a full breath-by-breath quantification of volatiles released from the mouth during mastication.

A typical result is shown in Fig. 3a.

One can see a very sharp leading edge, followed by an exponential decay as the flavour is depleted from the oil.
To obtain panel results, the individual curves are modelled by fitting a function consisting of two exponentials, one representing the rise and the other the decay of the signal, to the data.
This procedure transforms the individual breath-by-breath results into a smooth stimulus curve, characterised by the parameter values of the exponentials (see Fig. 3b).

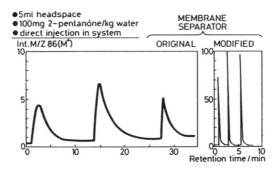

Fig. 1   Peak shape of headspace of 2-pentanone solution in water

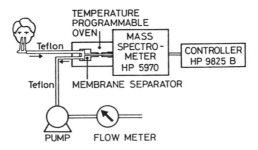

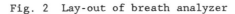

Fig. 2   Lay-out of breath analyzer

The parameter values are subsequently averaged to produce a panel stimulus curve. Apparently, the pentanone is depleted from the liquid layer in the mouth very rapidly. It should be borne in mind that this is not only due to release into the gas phase, but must also be ascribed to uptake into the mouth and upper airways (3).

Fig. 4 depicts the panel curves of the simultaneous release of butanone-2 and pentanone 2 from water. As pentanone-2 is more hydrophobic it is released faster.
Both curves peak at the same time but pentanone-2 peaks higher and is subsequently depleted faster. After about 50 s the release curves cross.
  If the flavour characters of these substances would have been sufficiently different the panel would probably have commented that butanone-2 released "late".

A time-dependent form of the psychophysical function

In order to be able to predict the effect of alterations to the time course of stimulation on perceived intensity over time, the static psychophysical function had to be extended.
  In the following section, taste and smell will be treated equally. In detail this is not correct but for the line of thought to be developed here the treatment is the same.

We started from Stevens' law (4) including the threshold correction:

$$I = k (S - So*)^n$$

where
T = perceived intensity as expressed
S = physical stimulus strength
So*= unadapted threshold level
k and n are constants

If prolonged stimulation, of any temporal form, is to have an effect on this relationship, i.e. if I becomes I(t), then at least one of the other parameters must also become a function of time. The only well documented effect of prolonged stimulation on the characteristics of taste and smell is adaptation.

Fig. 5a shows the effects of adaptation to a constant stimulus prior to magnitude estimation as measured by Cain (5); the curves relating Intensity to Stimulus strength drop off near the concentration level of the adapting stimulus. At higher stimulus levels, however, they seem to converge.
  In Fig. 5b the adapting levels have been deducted from the actual stimulation and results show straight lines for perceived intensity against stimulus minus adapting level. Within Stevens' equation, therefore, it appears that it is the threshold term that is affected by stimulation. We had to find out how, however.

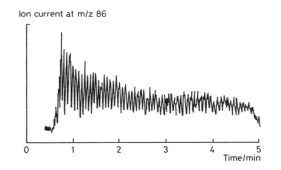

Fig. 3a Breath analysis
          100 mg 2-pentatone/kg MCT oil in the mouth

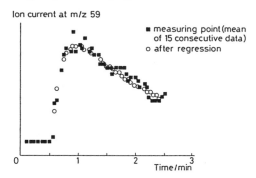

Fig. 3b Single release curve of butanol-2 from water after
        smoothing and after regression. Each black square
        represents the mean of 15 consecutive data points in a
        single experiment. The open circles represent the best
        fitting curve.

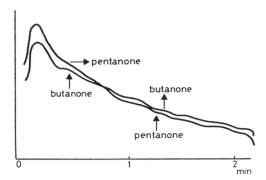

Fig. 4 Simultaneous release of butanone-2 and pentanone-2 from
       water

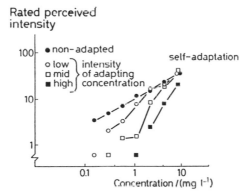

Fig. 5a Perceived intensity (I) vs stimulus concentration of
pentanol in an olfactometer experiment, under various
conditions of pre-adaptation   (after W.S. Cain, Percept.
Psychophys. 7 (1970) 271)

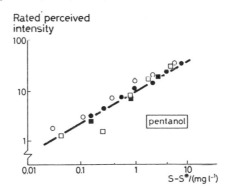

Fig. 5b Perceived intensity (I) of pentanol vs stimulus concentration (data from Fig. 5a) replotted after subtraction
of the adapting concentration from the stimulus concentration (I vs. S-S*)

In Fig. 6a data by Hahn (6) are shown. Hahn determined the effect of three levels of salt concentration, as a function of time, on the threshold of perception.

If we now define two extra parameters:

$S*$ = threshold level as a function of time
A = adaptation constant

the conclusions drawn from these data can be used to construct a differential equation relating threshold level to stimulation. The conclusions from the measurements by Cain and Hahn are:

- After prolonged stimulation the threshold rises to a level which lies above the level of stimulation, the difference being roughly equal to the original unadapted threshold level (for $t \to \infty$, $S* \to S + So*$).
- The adaptation proceeds faster when the difference between threshold and

stimulus is bigger; $\dfrac{dS*}{dt}$ $S - S*$

- The time it takes for the threshold to reach the stimulus level is longer

for a stronger stimulus; $\dfrac{dS*}{dt}$ $\dfrac{A}{S}$

Putting these conclusions together, we arrive at:

$$\frac{dS*}{dt} = \frac{A}{S} (So* + S - S*)$$

which can be integrated to give

$$S* = So* + e^{-\int \frac{A}{S} dt} \; . \; A \int e^{\int \frac{A}{S} dt} \, _{XL1} dt$$

in case of constant stimulation this reduces to

$$S* = So* + S (1 - e^{-\frac{At}{S}})$$

Fig. 6b shows the best fit of this equation to Hahn's data. Assuming that this relationship would also be valid for non-constant stimulation, we can try to predict what would happen if we would use a stimulus like the one we measured with the MS/breath method, after smoothing, at two levels of concentration (See Fig. 7).

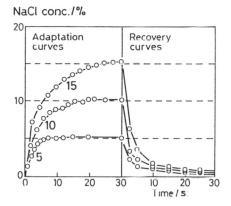

Fig. 6a Perception of thresholds vs. time under stimulation of 5, 10 and 15% sodium chloride solutions [after H. Hahn, Z. Sinnephysiol. 65 (1934) 105]

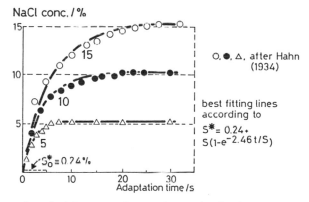

Fig. 6b Perception thresholds vs. time under stimulation of 5, 10 and 15% sodium chloride solutions

The theory predicts that in both cases the threshold level rises
linearly with stimulation in the first part of the curve and keeps
rising until the threshold level equals the level of stimulation.
Since we still use Stevens' law which has now taken a time-dependent
form: $I = k (S-S*)^n$, we may predict that the higher concentration
will be perceived

- more intensely
- at the same time of maximum intensity
- for a longer duration

Now that we have measured the actual stimulus shape and have
predicted its perceptual result, we evidently have to measure
perceived intensity as a function of time.

<u>Measuring perceived intensity over time (I/t)</u>

Methods for scoring perceived intensity over time are known in the
literature (7). They make use of a pen recorder, a dial
potentiometer or a "mouse" device coupled to a personal computer.
The panellists move the pen or dial up when perceived intensity
increases and down when it drops off. The data are pooled by
calculating mean Intensity values.
     The procedure contrasts with the above described MS/breath data
pooling method. In both cases we start with individual I/t curves.
In the MS/breath case these are parametrized, so that after pooling
the parameter values of the panel curve are the mean values of the
individual parameters.

The literature method for perceived intensity over time does not
produce such panel curves.
     Du Bois and Lee (8) describe a method which does produce panel
averages for the three main parameters: maximum perceived intensity
(I max) as scored by the individual panellists, the time at which
this occurs (t max) and the extinction time (t end). Since these
parameters do not produce a complete curve, we have developed a
method which produces complete curves which can be considered to be
real panel averages. This method is carried out as follows:

All individual curves are normalised in the Intensity direction by
calculating the geometric mean of all individual Imax values and
multiplying each

individual curve by $\dfrac{\text{Imax (geom)}}{I_i\text{max}}$

Subsequently all half curves before and after $t_i$max are averaged in
the time direction. Again the geometric mean is taken because a
check on the distribution of $t_i$max and t_i-end values (ATCS in Fig.
8) showed a log normal distribution. The resulting curve can be
considered to be a real panel average.

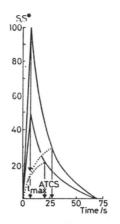

Fig. 7   Dependence of threshold of perception S* (•••) on an arbitrary time course of stimulation (___) at two levels

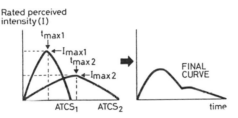

Fig. 8a Schematic representation of the existing procedure. The final curve is obtained after averaging the individual curves in the intensity direction only

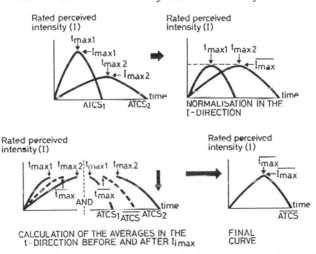

Fig. 8b Schematic representation of the new procedure. The final curve is obtained after averaging the individual curves in both the intensity and the time direction

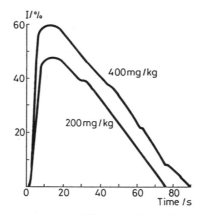

Fig. 9   Perceived intensity (I) vs. time of 2-pentanone in vegetable oil using the slide potentiometer method

Figs. 8a/b show a slightly simplified version of both types of data treatment mentioned. Our approach may be illustrated through the combination of two experiments.

The first involves I/t measurements of two concentrations of pentanonin in vegetable oil. The predicted results are obtained: a higher maximum and at the same time a longer duration for the higher concentration (see Fig. 9).

When these results are compared with those of the measurement of the real stimulus (Fig. 3a) the adaptation effect is evident; perception already ends when the actual stimulus has dropped only to around half of its highest value.

Summing up, we have defined our system as follows:

flavour/matrix > stimulus (t)   (psychophysical function (t) > response (t)

We have measured both time-dependent variables:

the actual stimulus and the response, and it has been shown that a suitable time-dependent version of Stevens' law could be constructed from material available in the literature (9, 10).

For stimuli containing more than one component it was shown that different physical release rates, starting at the same time, could very well give rise to perceived differences in release times.

References

1. R.M. Pangborne, Flavour 81 3-32, P. Schreier ed., 1981.

2. F.M. Benoit, W.R. Davidson, A.M. Lovett, S. Nacson, A. Ngo, Breat analysis by atmospheric pressure ionization mass spectrometry, Anal. Chem. 55, 805-807 (1983) and references therein.

3. H.K. Wilson and T.W. Ottley, The use of a transportable mass spectrometer for the direct measurements of industrial solvents in Breath, Biomedical Mass Spectrometry, 8 (12) (1981).

4. M. Stupfel and M. Mordelet-Dambrine, Penetration of pollutants in the airways. Bull. Physiopath. resp. 10, 481-509 (1974). A.H. Beckett and R.D. Hossie, Buccal absorption of drugs, Handbook of experimental Pharmacology, 28, 24-26 (1971).

5. S.S. Stevens, The surprising simplicity of sensory metrics. Am. Psychol., 17, 29-39, (1962).

6. W.S. Cain, Odor intensity after adaptation and cross adaptation, Percept. Psychophys. 7, 271-275 (1970).

7. H. Hahn, Die Adaptation des Geschmacksinnes. Z. Sinnesphysiol. 65, 105-145 (1934).

8. P. Overbosch, J.C. van den Enden and B.M. Keur, An improved method for measuring perceived intensity/time relationships in human taste and smell, Chemical Senses, 11, (3) pp. 331-338 (1986).

9. G.E. DuBois and J.F. Lee, A simple technique for the evaluation of temporal taste properties, Chem. Senses, 7, 237-247 (1983).

10. P. Overbosch, A theoretical model for perceived intensity in human taste and smell as a function of time, Chemical Senses, 11, (3) pp. 315-329 (1986).

11. W.J. Soeting and J. Heidema: A mass spectrometric method for measuring flavour concentration/time profiles in human breath. To be submitted for publication.

RECEIVED September 23, 1988

Chapter 12

Enantioselectivity in Odor Perception

W. Pickenhagen

Research Laboratories, Firmenich SA, Case Postale 239, CH−1211, Geneva, Switzerland

The first molecular event in odor perception is an interaction of an odorant with a receptor. Evidence exists that these receptors are proteins, i.e. chiral, so this first interaction should be enantioselective, meaning that these receptors react differently with the two enantiomeric forms of a chiral odorant leading to differences in odor strength and quality. In many cases, this fact has been observed.This paper describes the enantioselective syntheses of some known odorants of multiple chemical classes and discusses the differences of the organoleptic properties of their enantiomeric forms.

The mechanism of odor perception is very complicated and the least understood of all our senses. It is well accepted that the perception of an odor, meaning the actual recognition by the brain, goes through a cascade of events.

$$\text{STIMULUS} \;\rightarrow\; \text{RECEPTOR} \;\rightarrow\; \text{TRANSDUCTION} \;\rightarrow\; \text{PROCESSING}$$

Of all these different steps, the very first one, namely the interaction of a stimulus, i.e. molecules that "have a smell", with the actual receptor is not at all understood. These receptors are supposed to be located in the membrane of the cilia cells, because these cilia are the furthest out of the antennae of the olfactory system, and they have been shown to be excitable by chemical stimuli.

In analogy to other – better understood – receptor systems like some hormone and opiate receptors it is generally accepted that the olfactory receptors are proteins, and there are some facts known that support this hypothesis. One of these arguments is that, sometimes, slight modification of the chemical structure of a stimulus molecule can lead to big changes in the odor impression; this might be qualitative or quantitative.

Proteins are chiral, so they should interact differently with the two enantiomeric forms of a chiral molecule, which should eventually translate into a difference of the odor impression of these mirror images of the molecules. A more detailed knowledge of the relations between the chemical structure of a molecule, including its absolute configuration, and its odor properties will contribute to the elucidation of the receptor mechanism.

Actually there are many examples known where the two enantiomeric forms of chiral compounds have different odors. Table I shows some of them without being exhaustive.

Enantioselective synthesis have become very fashionable in preparative chemistry, and a considerable effort is devoted to their methodology.

0097–6156/89/0388–0151$06.00/0
© 1989 American Chemical Society

Table 1
Enantiomeric Forms of Chiral Compounds and Their Odors

Compound		Odor impression	Lit.
	(+)-Linalool	sweet, petitgrain	[1]
	(–)-Linalool	lavender notes, Ho oil, woody	
	(+)-Carvone	caraway	[2]
	(-)-Carvone	spearmint	
	(+) -cis-Rose oxide	sweet	[3]
	(-) -cis-Rose oxide	powerful, fruity	
	(+)-Hydroxycitronellal	sweet, powerful	[4]
	(-)-Hydroxycitronellal	minty	

Table 1. *Continued*

Compound	Odor impression	Lit.
(+)-Nootkatone	t = 0.8 ppm grapefruit, strong	[5]
(−)-Nootkatone	t = 600 ppm very weak, no grapefruit	
(-)-Patchoulol	natural patchouli, earthy, cellary	[6]
(+)-Patchoulol	weak, not reminiscent of patchouli	
(-)-Androstenone	sweaty, urine musky, strong	[7]
(+)-Androstenone	odorless	[8]
(+)- *cis*-2-methyl-4 propyl-1,3-oxathiane	t = 2 ppb sulfury, rubbery, tropical fruit	[9]
(-)- *cis*-2-methyl-4- propyl-1,3-oxathiane	t = 4 ppb flat, estery, camphoracious	[10]

In our continuous interest in the relation of molecular structure and organoleptic activity, we synthesized the enantiomers of some well-known aroma chemicals, and evaluated their odor. For the preparation different synthetic approaches were used, i.e.

a)   starting with the same material employing reagents of opposite chirality that can either be recovered after use or are lost during the synthesis;

b)   starting with natural products of known antipodal configuration.

Muscone 1 was discovered in 1906 [11]; its structure [12] and absolute configuration [13] were determined later to be (R)-3-methylcyclopentadecanone.

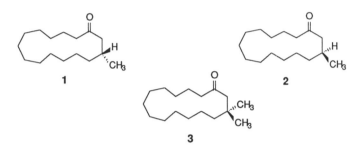

1   2

3

It is, in its racemic form, a highly appreciated ingredient in fine perfumery. Because of its value a number of syntheses have been described [14]. Enantioselective syntheses of the (–)-(R) 1 and the (+)-(S)-form 2 have been developed [15], however no olfactive description of the two compounds could be found. Following the synthesis of Nelson and Mash, both enantiomeric forms of muscone were prepared. The optical purity, determined by 360 MHz NMR, using Pr (hfbc)$_3$ as chiral shift agent was 95.5% for the (-)-(R) and 97.7% for the (+)-(S)-form. The two products show distinct differences in their odor. The natural 1 is described by a panel of perfumers as "very nice musky note, rich and powerful", whereas 2 is "poor and less strong". Thresholds, determined in water, using Guadagni's procedure [16], with a panel of 18 - 20 members, show values of 61 and 233 ppb respectively, giving a calculated threshold of 97 ppb for the racemic mixture, in good accordance with the experimental value of 103 ppb.

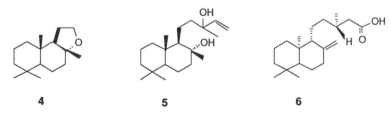

4   5   6

From these results, one could deduce that the methyl group in 2 somehow hinders easy access of the molecule to its receptor. This hypothesis is supported by the fact that 3,3-dimethylcyclopentadecanone 3 is nearly odorless.

The tricyclic ether AMBROX 4, first synthesized in 1950 [17], was later found as a constituent of ambergris [18], oriental tobacco (Demole, E. , Firmenich SA, unpublished data), clary sage (Renold, W.; Keller, U.; Ohloff,G. Firmenich SA, unpublished data) and ciste labdanum (Renold, W.; Wuffli, F.; Ohloff, G. Firmenich SA, unpublished data). The absolute configuration of the natural (-)-form is determined by the configuration of the starting material (-)-sclareol 5.

For the synthesis of the (+) enantiomer, eperuric acid 6 extracted from Wallaba wood *(Eperura falcata)* was converted to 7 following the method of Dey and Wolf [19]. The ketone 7 was then transformed into (+) Ambrox *ent*-4, following scheme 1 [20]. Optical purity, determined by 360 Mhz HNMR using Eu (hfbc)$_3$ as chiral shift agent, is more than 98%, confirmed also by capillary gas chromatography using Ni(hfbc)$_2$ in OV 101 as chiral stationary phase.

Scheme 1

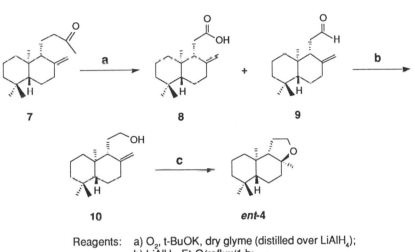

7   8   9

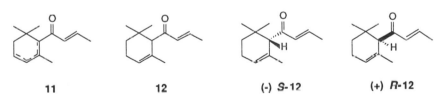

10   *ent*-4

Reagents: a) O$_2$, t-BuOK, dry glyme (distilled over LiAlH$_4$);
b) LiAlH$_4$, Et$_2$O/reflux/1 h;
c) CH$_3$NO$_2$, TsOH/reflux/100°/1 h.

Organoleptic comparison of the two forms shows that the (+) enantiomer has a dominant woody note and lacks the warm animal note of the (-)-form. Thresholds in water [16] were measured to 0.3 ppb for (-) 4 and 2.6 ppb for the (+). The racemic mixture was determined to be 0.6 ppb, corresponding well to the calculated threshold of 0.54 ppb.

11   12   (-) *S*-12   (+) *R*-12

The rose ketones 11, first discovered in 1970 [21] in Bulgarian rose oil, and named damascones, show unique organoleptic properties. Because of this they have elicited great interest, also as target molecules for new synthetic methods. α-Damascone 12 possesses a quite unique fruity odor, and its utilization allows the creation of perfumistic notes otherwise difficult to achieve.

Treatment of (+)-epoxy-α-dihydroionone with hydrazine hydrate gives as one of the reaction products alcohol 13, which was transformed by oxidation with MnO$_2$ to (+)-*(R)*-α-damascone 12 in 65% e. e., thus establishing its absolute configuration [22].

OH

13

A new access to α-damascone by selective kinetic protonation of α-ketone enolate, formed by reaction of an ester enolate with nucleophiles, has recently been described by Fehr and Galindo [23] (scheme 2).

Scheme 2

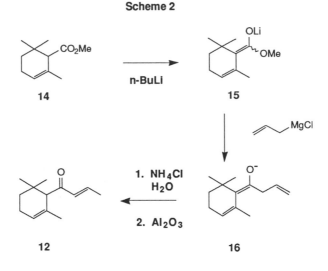

The same authors found that the prochiral enolate **16** can, under certain conditions, be protonated enantioselectively, using ephedrine derivatives as proton sources [23]. These compounds are available in their optically pure forms, thus both enantiomers of α-damascone can be prepared in about 70% optical yield starting with the same ketone enolate and using the appropriate optical form of the proton source. Enantiomerically pure α-damascones (-)-(S)- 12, (+)-(R)-12 have been obtained by repeated recrystallization.

The organoleptic properties of the two compounds are distinct. Striking is the difference in perception thresholds, which were found to be 1.5 ppb for the (-)-(S)-, and 100 ppb for the (+)-(R)-form. Qualitatively, the (-)-(S) is described as more floral, reminiscent of rose petals, also having a winy character without the "cork" and the green apple note that are the characteristics of the (+)-(R)-form as well as of the racemic mixture.

These examples that add to the existing list show to what extent modification of the chemical structure of a molecule can alter the perceived odor. The fact that two enantiomeric forms of odorants show distinct differences in their organoleptic properties supports the hypothesis that the initial event, the interaction of the stimulus with the receptor is enantioselective, leading to diastereoisomeric stimulus-receptor complexes; and these events are transduced to give rise to different odor impressions, the mechanism of which remains to be discovered.

Literature Cited

1. Ohloff, G.; Klein, E. Tetrahedron, 1981, 18, 37.
2. a) Friedmann, L.; Müller, J.G. Science, 1971, 172, 1044.
   b) Russel, G.F.; Hills, J.I. Science, 1971, 172, 1043.
   c) Leitereg, T.J.; Guadagni, D.G.; Harris, J.; Mon, T.R.; Teranishi, R. Nature, 1971, 230, 455.
   d) Leitereg, T.J.; Guadagni, D.G.; Harris, J.; Mon, T.R.; Teranishi, R. J. Agric. Food Chem, 1971, 19, 785.

3. Ohloff, G. In Olfaction & Taste IV; D. Schneider, Ed.; Wiss. Verlagsges.: Stuttgart, 1972, p 156.
4. Skorianetz, W.; Giger, H.; Ohloff, G. Helv. Chim. Acta, 1971, 54, 1797.
5. Haring, H.G.; Rijkens, F.; Boelens, H. v. d. Gen A. J. Agric. Food Chem., 1972, 20, 1018.
6. Näf, F.; Decorzant, R.; Giersch; W.; Ohloff,G. Helv. Chim. Acta, 1986, 64, 1387.
7. Prelog, V.; Ruzicka, L.; Wieland, P. Helv. Chim. Acta, 1944, 27, 66.
8. Ohloff, G.; Maurer, B.; Winter, B.; Giersch, W. Helv. Chim. Acta, 1983, 66, 192.
9. Pickenhagen, W.; Brönner-Schindler, H. Helv. Chim. Acta, 1984, 67, 947.
10. Heusinger, G.; Mosandl, A. Liebigs Ann. Chem. 1985, 1185.
11 Walbaum, H. J.Prakt. Chem. II, 1906, 73, 488.
12. Ruzicka, L. Helv. Chim. Acta, 1926, 9, 715, 1008.
13. Ställberg-Stenhagen, S. Ark. Kemi, 1951, 3, 517.
14. For a review see Mookherjee, B.; and Wilson, R.A. In Fragrance Chemistry; Theimer, E.T., Ed.; Academic Press Inc.: New York, 1982, p 433; and Wood, T.F. In Chemistry of Synthetic Musks, Ibid. p 495.
15. Branca, Q.; Fischli, A. Helv. Chim. Acta, 1977, 60, 925. Nelson, K.A.; Mash, E.A., J. Org. Chem., 1986, 33, 2171.
16. Schwimmer, J.; Guadagni, D.G. J. Food Sci., 1962, 27, 94.
17. Hinder, M.; Stoll, M. Helv. Chim. Acta, 1950, 33, 1308.
18. Mookherjee, D.D., Patel, R.R. Proc. of the VIIth Int. Cong. of Ess. Oils, 1977, p 479.
19. Dey, A.K.; Wolf, H.R. Helv. Chim. Acta, 1978, 61, 1004.
20. Ohloff, G.; Giersch, W.; Pickenhagen, W.; Furrer, A.; Frei, B. Helv. Chim. Acta, 1985, 68, 2022.
21. Demole, E.; Enggist, P.; Säuberli; U.; Stoll, M. Helv. Chim. Acta, 1970, 55, 541.
22. Ohloff, G.; Uhde, G. Helv. Chim. Acta, 1970, 55, 531.
23. Fehr, C.; Galindo, J. J. Org. Chem., 1988, 53, 1828.
24. Fehr, C.; Galindo, J. Submitted to J. Am. Chem. Soc.

RECEIVED September 12, 1988

Chapter 13

Role of Free Amino Acids and Peptides in Food Taste

Hiromichi Kato, Mee Ra Rhue, and Toshihide Nishimura

Department of Agricultural Chemistry, University of Tokyo, Bunkyo-Ku, Tokyo, Japan

Free amino acids and/or some peptides have some
sweetness, bitterness, sourness, saltiness and umami,
and are very important as taste substances in foods.
In this paper, we discuss 1)some tastes of free amino
acids and some peptides, 2)the role of free amino acids
in the characteristic tastes of vegetables and marine
foods, 3)the role of the bitter peptides in cheese and
the traditional Japanese foods "miso" and "natto", and
4)the contribution of free amino acids and peptides to
the improvement of the meat taste during storage of
meats (beef, pork and chicken).

Free amino acids and peptides are very important as taste
substances. The importance of amino acids to food taste was first
recognized by Ikeda in 1908 (1). He discovered that monosodium
glutamate (MSG) was the essential taste component of traditional
Japanese seasoners, such as sea tangle. MSG is a typical umami
substance. Almost all free amino acids, including MSG, have some
sweetness, bitterness, sourness and umami (2,3) and therefore contribute to the characteristic taste of foods. The characteristic
taste of many marine foods is elicited by free amino acids. The
taste of traditional Japanese foods such as sake, miso and soy sauce
is thought to be caused by amino acids released from proteins during
fermentation. Many studies on the taste of amino acids in foods and
their production by extraction, fermentation, or chemical syntheses
have resulted in mass production of various amino acids. Today,
amino acids used in food processing not only enhance the nutritive
value of many processed foods such as cereals, but also enhance the
natural characteristic tastes of many foods.
Studies on the taste of peptides have been done only recently.
The bitter taste produced during the storage of cheese and in the
fermentation of the traditional Japanese food "miso" and "soy sauce"
has been shown to be caused by the peptides in the hydrolysate of
proteins. Since then, a number of studies on bitter peptides have

0097–6156/89/0388–0158$06.00/0
© 1989 American Chemical Society

been conducted. Also, research has been conducted on sweet and umami peptides recently.

This paper deals with the tastes of free amino acids and peptides, and their roles in the taste of foods.

For the convenience of our readers, the amino acid and peptide symbols used are:

Alanine	Ala	Leucine	Leu
Arginine	Arg	Lysine	Lys
Asparagine	Asn	Methionine	Met
Aspartic acid	Asp	Phenylalanine	Phe
Asn + Asp	Asx	Proline	Pro
Cysteine	Cys	Serine	Ser
Glutamine	Gln	Threonine	Thr
Glutamic acid	Glu	Tryptophan	Trp
Gln + Glu	Glx	Tyrosine	Tyr
Glycine	Gly	Valine	Val
Histidine	His	Anserine	Ans
Isoleucine	Ile	Carnosine	Car

TASTE OF FREE AMINO ACIDS

Amino acids are not only the building blocks of proteins but also occur in the free form. Amino acids commonly found in proteins have the L-configuration. Of these amino acids (Table 1), Asn was first discovered in asparagus in 1806, and Thr, the most recently discovered, was isolated from the hydrolysates of fibrin in 1935. Most of them were isolated from hydrolysates of various proteins. Glu, first obtained from wheat gluten hydrolysate in 1886, was found to be the most important taste component in sea tangle by Ikeda in 1908. Later, industrial production of MSG was undertaken to utilize it as a seasoner.

Almost all amino acids elicit taste. Most hydrophobic L-amino acids have a bitter taste. However, hydrophobic D-amino acids, which are formed simultaneously by the synthesis of L-amino acids, bring out a strong sweet taste. D-Trp, Phe, His, Tyr and Leu are 35, 7, 7, 6 and 4 times as sweet as sucrose, respectively (2). Gly and L-Ala elicit a strong sweet taste. It is thought that the strong sweet taste elicited by these amino acids is due to the ability of these molecules to bind to the sweet substance receptors.

L-Glu and Asp are sour stimuli in dissociated state, but their sodium salts dissociate on solution and elicit the umami taste. Free L-glutamate is contained in natural foods, as shown in Table 2 and contributes to the savory taste of foods as its sodium salt. Ibotenic and tricholomic acids (IA and TA) discovered in mushrooms are the derivatives of oxyglutamic acid and are also umami substances (4,5). The umami taste intensity of IA or TA is 4 to 25 times that of MSG. As these compounds are not amino acids commonly found in an animal system, they have not been used as seasoners. The umami taste of a MSG-, IA- or TA-5'-ribonucleotide mixture is much more intense than that of only MSG, IA or TA. Among 5'-ribonucleotides, 5'-inosinate and guanylate have synergistic effects in a mixture with MSG, IA or TA. This phenomenon is called the synergistic effect of

Table 1. Taste, Threshold Value and Discovery of Amino Acids

Amino acid	Taste	Threshold value(mg/dl)	Where found
His	Bitter	20	Casein and sturin*(1896)
Met	Bitter	30	Casein*(1922)
Val	Bitter	40	Albumin*(1879)
Arg	Bitter	50	Casein*(1895)
Ile	Bitter	90	Syrup(1904)
Phe	Bitter	90	Bean sprouts(1881)
Trp	Bitter	90	Casein*(1901)
Leu	Bitter	190	unknown(1819)
Tyr	Bitter	ND	Casein*(1846)
Ala	Sweet	60	Fibroin*(1875)
Gly	Sweet	130	Gelatin*(1820)
Ser	Sweet	150	Sericin*(1865)
Thr	Sweet	260	Fibrin*(1935)
Lys	Sweet and bitter	50	Casein*(1889)
Pro	Sweet and bitter	300	Casein*(1901)
Asp	Sour	3	Asparagine*(1827)
Glu	Sour	5	Gluten*(1886)
Asn	Sour	100	Asparagus(1806)
Gln	Flat		Beet(1883)
Cys		ND	Cystine(1884)
Glu Na	Umami	30	Sea tangle(1908)
Asp Na	Umami	100	unknown

ND, not determined; *, hydrolysate.

taste (6). When Gly was added to a MSG-5'-ribonucleotide mixture, the umami taste intensity of the mixture was greater than that of the mixture before addition (7). Ala, Cys, His, Met, Pro and Val, besides Gly, were also recognized as having the synergistic effect of taste in the mixture of MSG, 5'-ribonucleotide and free amino acids (8).

Though amino acids can elicit any one of the primary tastes, the threshold value of taste of each amino acid is high. As the levels of some free amino acids in natural foods are lower than their threshold values, it may be thought that they may not contribute directly to food taste. However, they may have an important role in making the food savory because of the synergistic effect.

TASTE OF PEPTIDES

Sweet Taste
        The sweet peptide, aspartame (L-Asp-L-Phe-OMe) which has a sweet taste 180 times that of sucrose, was discovered by Mazur et al. (9). Aspartame is stable at pH 4 and unstable at pH 1 or 7-8. It is also unstable at high temperatures. Under these unstable conditions, the

Table 2. Free L-glutamate in Natural Foods

Food	L-glutamate (mg/100 g)
Kelp	2240
Parmesan cheese	1200
Green tea	668
Seaweed	640
Fresh sardine	280
Fresh tomato juice	260
Champignon	180
Tomato	140
Oyster	137
Potato	102
Chinese cabbage	100
Fresh shiitake mushroom	67
Soybean	66
Sweet potato	60
Dried sardine	50
Prawn	43
Clam	41
Chicken bones	40
Cabbage	37
Carrot	33
Bonito flakes	26
Pork fillet	23

ester linkage of aspartame may hydrolyze to produce aspartyl-phenylalanine (AP) or cyclize to produce the corresponding diketo-piperazine (DKP). As none of these conversion products fits the sweet taste receptor, none of them is sweet.

Thaumatin (10) and monellin (11) are sweet and are proteins isolated from a plant native to Western Africa. Thaumatin and monellin are respectively 1600 and 3000 times sweeter than sucrose. As both proteins are basic, they are assumed to bind easily to the negatively charged taste cell.

There have been no reports regarding the detection of sweet peptides in naturally occurring foodstuffs other than thaumatin and monellin.

Bitter Taste

Almost all peptides of hydrophobic L-amino acids elicit a bitter taste, which indicates that the bitterness of peptides is caused by the hydrophobic property of the amino acid side chain. Ney (12) has reported that whether a peptide has a bitter taste or not depends on its hydrophobic value Q. The value Q is obtained by adding the Δf-values (Table 3) of each constituent amino acid residue of a peptide and dividing the sum by the number of amino acid residues (n).

$$Q = \frac{\sum \Delta f}{n}$$

If the value Q of a peptide is greater than 1400, the peptide

Table 3. Δf-value of the side chain of amino acid

amino acid	Δf (cal/mol)
Gly	0
Ala	730
Val	1690
Leu	2420
Ile	2970
Phe	2650
Pro	2600
Met	1300
Tyr	2870
Thr	440
Ser	40
Asp	540
Glu	550
Arg	730
Lys	1500
Trp	3000

will elicit bitter taste. This rule is applicable to almost all peptides.

Hydrolysis of proteins without taste by proteases often produces bitter peptides. Hydrophobic amino acid residues located in the interior of protein molecules in aqueous solution are exposed by fragmentation of the protein molecules treated with proteases, and the peptides containing a number of hydrophobic amino acid residues occur in the solution (13). Many bitter peptides as shown in Table 4 have been isolated from protein digests with proteinases (14-22).

The comparison of the amino acid sequence of the above-mentioned bitter peptides shows a large proportion of hydrophobic amino acids in each peptide. And the amino acid sequence of peptides also plays an important role in the intensity of the bitter taste. For example, the bitterness of Phe-Pro is more intense than that of Pro-Phe, and the bitterness of Gly-Phe-Pro is more intense than that of Phe-Pro-Gly (23). C-terminal groups of all bitter peptides in pepsin hydrolysates of the above-mentioned soy protein were characterized by the location of the Leu residue (14-17). The research on the relationship between the structure and bitter taste intensity of Arg-Gly-Pro-Pro-Phe-Ile-Val (BP-Ia) showed that Pro and Arg located on center and the N-terminal site, respectively, played an important role in the increment of bitter taste intensity besides the hydrophobic amino acids located on C-terminal site (24-26). This may indicate that the peptide molecular structure formed by the arrangement of Arg, Pro and hydrophobic amino acid residues contributes to the bitter taste intensity of the peptide.

Sour Taste

As shown in Table 5, dipeptides containing Glu and/or Asp, Gly-Asp-Ser-Gly, Pro-Gly-Gly-Glu and Val-Val-Glu in water elicit

Table 4. Bitter Peptides Isolated from Proteinase Hydrolysates of Proteins

Protein (proteinase[x])	Peptides isolated from hydrolysate
Soy protein (pepsin)	Gly-Leu, Leu-Phe, Leu-Lys, Arg-Leu, Arg-Leu-Leu, Ser-Lys-Gly-Leu, PyroGlu-Gly-Ser-Ala-Ile-Phe-Val-Leu, Tyr-Phe-Leu, Phe-Leu
Zein (pepsin)	Ala-Ile-Ala, Ala-Ala-Leu, Gly-Ala-Leu, Leu-Gln-Leu-Leu-Glu-Leu, Leu-Val-Leu, Leu-Pro-Phe-Asn-Gln-Leu, Leu-Pro-Phe-Ser-Gln-Leu
Casein (papain)	Ala-Gln-Thr-Gln-Ser-Leu-Val-Tyr-Pro-Phe-Pro-Gly-Pro-Ile-Pro-Asn-Ser-Leu-Pro-Gln-Asn-Ile-Pro-Pro-Leu-Thr-Gln
Casein (trypsin)	Gly-Pro-Phe-Pro-Val-Ile, Phe-Phe-Val-Ala-Pro-Pro-Glu-Val-Phe-Gly-Lys, Phe-Ala-Leu-Pro-Glu-Tyr-Leu-Lys
Casein (bacterial proteinase)	Arg-Gly-Pro-Pro-Phe-Ile-Val, Val-Tyr-Pro-Phe-Pro-Pro-Gly-Ile-Asn-His, cyclo(Leu-Trp-Leu-Trp)

*, used to obtain hydrolysate of protein.

sour taste ($\underline{3}$). This acidic sensation is assumed to be produced by the binding of the hydronium ion, produced by the dissociation of acidic amino acid, to the taste cell membrane.

Table 5. Sour Peptides

Gly-Asp, Gly-Glu; Ala-Asp, Ala-Glu; Ser-Asp, Ser-Glu; Val-Asp, Val-Glu; Asp-Ala, Asp-Asp; Glu-Ala, Glu-Asp, Glu-Glu; Glu-Phe[a], Glu-Tyr[a], γ-Glu-Gly[b], γ-Glu-Ala[b], γ-Glu-Asp[b], γ-Glu-Glu[b]; Phe-Asp, Phe-Glu, Trp-Asp, Trp-Glu; Gly-Asp-Ser-Gly, Pro-Gly-Gly-Glu, Val-Val-Glu

All amino acids have the L-configuration. a, Accompanied by bitterness and astringency; b, Accompanied by astringency.

Salty Taste

It has been reported that there are salty stimuli in peptides. Tada et al. ($\underline{27}$) inadvertently discovered the synthesized salty dipeptides, L-Orn-β-Ala·HCl, L-Orn-Tau · HCl, Lys-Tau·HCl and L-Orn-Gly · HCl having the same intensity taste as NaCl. The salty taste of L-Orn-Tau · HCl and Lys-Tau·HCl was more intense than that of L-Orn-β-Ala · HCl and L-Orn-Gly · HCl. The degree of dissociation of the carboxyl or sulfonyl group in peptides was assumed to contribute to the intensity of the salty taste. These dipeptides may be useful as new seasonings for diabetics and hypertensives because they contain no Na ions.

Recently, Huynh-ba and Philippossian (28) have reported that the L-Orn-Tau · HCl, L-Orn-β-Ala·HCl and L-Orn-Gly·HCl they synthesized elicited no salty taste. The salty taste of L-Orn-Tau · HCl synthesized by Tada et al. seemed to result from the NaCl present as an artifact in the method of preparation. However, the preparation of L-Orn- β -Ala·HCl (OBA·HCl) and L-Orn-Gly·HCl did not contain NaCl at all. The reason for this discrepancy is not clear yet. We heard from Okai group that the purified OBA without HCl did not elicit salty taste in water. When the molar ratio of HCl to OBA in OBA·HCl became 0.97 in the solution, this solution elicited a slightly salty taste. This salty taste elicited strongly with increasing HCl, till the molar ratio of HCl to OBA became 1.3 (Okai, H., Hiroshima University, personal communication, 1988.). As the molar ratio of HCl in OBA·HCl synthesized by Huynh-ba and Philippossian was 1.0, its salty taste might be very weak.

Umami Taste

Several dipeptides having L-Glu at N-terminus elicit the umami taste, though its umami taste intensity is much less than that of MSG. Arai et al. (29) synthesized L-Glu-X (X= amino acid) and examined their taste in aqueous solution containing NaCl at pH 6. Glu-Asp, Glu-Thr, Glu-Ser and Glu-Glu were found to produce the umami taste. Ohyama et al. (30) showed that Asp-Leu and Glu-Leu were umami substances. In section "Sour Taste", the peptides containing Asp or/and Glu were shown to elicit a sour taste in water. However, several of their peptides besides Glu-Asp and Glu-Glu may also be umami stimuli in aqueous solutions containing NaCl at pH 6.

When fish proteins were thoroughly hydrolyzed by pronase, the hydrolysate elicited the complex taste containing bitterness. Peptides having not only the bitter but also the umami taste were produced in this hydrolysate. The fraction of compounds with molecular weight under 500 was obtained from this hydrolysate by ultrafiltration. This fraction was divided into four fractions, aromatic, acidic, neutral and basic fractions. The acidic fraction had a very intense umami taste. Though Glu was removed from the acidic fraction by the treatment with ion-exchange chromatography, the treated acidic fraction also elicited an umami taste. Therefore, the umami taste of the acidic fraction can be ascribed in part to peptides, although MSG was mainly responsible for the umami taste. Umami peptides from this fraction (Table 6) were characterized by containing Glu residue and a number of hydrophilic amino acids except for Glu (31,32).

Table 6. Umami Peptides Isolated from Fish Protein Hydrolysates by Pronase*

Dipeptides:	Glu-Glu, Glu-Asp, Thr-Glu, Glu-Ser
Tripeptides:	Glu-Gly-Ser, Ser-Glu-Glu, Glu-Gln-Glu, Glu-Asp-Glu, Asp-Glu-Ser

*, Threshold values of these peptides are 150-300 mg/dl.

An octapeptide, Lys-Gly-Asp-Glu-Glu-Ser-Leu-Ala, which has a delicious taste, was isolated from beef treated with papain by the use of gel filtration and ion-exchange chromatography, and filter paper electrophoresis (33). A sensory evaluation showed that this octapeptide produced synthetically also elicited a delicious taste (34). The elimination of two amino acid residues at N-terminus, Lys-Gly, led to the disappearance of the savory taste and changed it into a sour taste. This indicated that the residues, Lys-Gly are important to the savory taste.

Other Tastes

Kirimura et al. (3) have reported that the dipeptides formed by the binding of γ-COOH group in Glu to NH₂ group in another amino acid, such as γ-L-Glu-Gly, γ-L-Glu-L-Ala, γ-L-Glu-L-Asp and γ-L-Glu-L-Glu, have not only sour taste but also astringent taste.

In general, because peptides were amphoteric electrolytes, they have a buffer action on taste. β-Ala-His (Car) and β-Ala-1-methyl-His (Ans) widely distributed in animal tissues were found to have a large buffer action in the pH range above 6.0 (35). Dipeptides, Gly-Leu, Pro-Glu and Val-Glu, were also found to have a buffer action (3). When these peptides were added to a synthetic "sake", a traditional Japanese alcoholic drink, composed of alcohol, glucose, succinic acid, lactic acid, phosphoric acid, NaCl, MSG, Gly and Ala, the buffer actions of sake containing Gly-Leu and Pro-Glu were larger in the pH range above pH 8 than that of sake without Gly-Leu adn Pro-Glu. The buffer action of sake containing Val-Glu was larger in the pH range 7-9 than that of sake not containing Val-Glu. The buffer action seems to play an important role in the improvement of food taste by enhancing the taste of food and keeping the elicitation of its taste (36).

A glutamic acid-rich oligopeptides fraction was found to be effective in masking bitter taste (37). The addition of these peptides to the bitter medicinal drugs and drinks (summer orange and vegetable juices, and cocoa) seems to decrease or mask the bitterness of the products.

ROLE OF FREE AMINO ACIDS AND PEPTIDES IN FOOD TASTES

Vegetable Foods

Free amino acids play an important role in the taste of vegetables. There are large amounts of Glu, Asp, Ser, Val, Ala, Pro and Gln in vegetables as shown in Table 7 (38). The detailed research on the taste of green tea, onion and potato reveals the presence of umami substances. The most important umami substances of green tea are Glu and L-theanine, which is an ethylamide derivative of Glu (39-41). It has been shown that the most important umami substance of onion (42) is Glu. Buri et al. (43) examined the role of free amino acids in the flavor of boiled potatoes. Analytical data has shown that the taste of boiled potato soup stock was similar to that of synthetic potato soup composed of free amino acids and nucleotides. This indicated that free amino acids are very important in potato taste. Although they contain large amounts of Glu and Asp, there are

Table 7. Contents of Free Amino Acids in Vegetables
(α-amino N mg/100 g)

Amino acid	Tomato	Egg plant	Cucumber	Carrot	Pumpkin	Maize
Glu	3.99	0.84	0.65	3.02	3.03	0.33
Ser	10.07	0.59	2.85	-	2.28	0.55
Gly	4.83	0.48	0.54	-	0.82	0.44
Asn	6.61	2.66	3.33	2.34	9.77	1.32
Lys	1.89	-	-	-	1.16	-
Thr	0.11	+	0.35	0.33	0.46	0.22
Gln	6.58	3.02	-	1.44	5.09	4.18
Ala	0.92	0.97	0.95	1.33	2.24	1.32
Arg	6.81	2.05	0.53	4.11	1.95	1.32
Tyr	0.75	0.34	0.65	0.12	0.45	0.11
Val	9.17	2.35	1.10	1.23	4.01	3.52
Phe	2.37	-	0.65	-	2.05	0.33
Leu	2.14	0.95	0.95	0.45	0.74	0.77
Pro	-	0.54	0.50	0.55	1.71	1.76
Asp	-	0.99	0.80	2.88	-	0.66
Cys	-	0.50	-	-	-	+

+, trace; -, not detected.

Table 8. 5'-guanylate in Natural Foods

Food	5'-guanylate (mg/100 g)
Dried shiitake mushroom	156.5
Matsutake	64.6
Enokitake mushroom	21.8
Fresh shiitake mushroom	16-45
Truffle mushroom	5.8
Pork	2.5
Beef	2.2
Chicken	1.5

smaller amounts of nucleotides, IMP or GMP, in potatoes than in animal foods. In animal foods, IMP or GMP enhances umami and brothy taste elicited by MSG. This may account for the absence of the brothy taste in vegetables (44). However, mushrooms contain exceptionally high levels of GMP (Table 8). GMP in mushrooms enhances the umami taste of Glu by a synergistic effect and imparts a brothy taste.

Miso and natto are traditional Japanese foods made from soybeans by a fermentation process. These foods are produced by a mixed fermentation process using a characteristic microorganism and ripening for a given time. In these processes, the hydrolysis of proteins by microbial proteases results in the production of free

amino acids and peptides. Free amino acids and peptides produced in
each food contribute to the characteristic taste of each food. The
rate of liberation of Glu and Asp as umami substances during ripening
of miso was very slow and amounts liberated from proteins were small
(45). Examination of the change in peptides during storage of miso
showed that the peptides of A.P.L.(the average number of amino acid
residues in the peptide) 3-4 gradually decreased, but the peptides of
A.P.L. 3-4 rapidly increased during the initial stages of storage.
The peptides of A.P.L. 13-20 increased with storage time (Fig. 1)
(46). As 40 % of the constituent amino acids residue in the peptides
of A.P.L. 13-20 was Glu, these peptides seemed to play an important
role in the umami taste of miso.

The free amino acid content in natto was very small and
corresponded to only about 10 % of the total nitrogen compounds. Most
of the other nitrogen compounds were peptides. These peptides have
been shown to contribute to the bitterness of the characteristic
taste of natto (47). One of these peptides was isolated and its
amino acid sequence was investigated. The amino acid composition of
this peptide was Asp 1, Thr 1, Glu 1, Ala 1, Pro 2, Val 3, Ile 3 and
Leu 5. The amino acid at N-terminus of this peptide was Leu and the
C-terminal structure was -Ala-Val-Ile-Leu.

A cyclic dipeptide, Pro-Leu anhydride, having bitterness was
isolated from a traditional Japanese alcoholic drink "sake" (48).
This peptide increased the longer sake was stored in sake production.
So this peptide seems to contribute to the bitter taste of sake.

Animal Foods

Taste components of a number of sea food products have been
examined for each sea food product has its individual characteristic
taste. In studies on the free amino acids analysis (Table 9), it was
shown that the major amino acid is His in red meat of fish, Gly and
Pro in cuttlefish, Gly and Arg in prawns, and Tau and Arg in abalone
(49). The major amino acids in sea urchin are Gly, Ala and Leu (50).
However, the components contributing to their individual characteristic tastes were not elucidated because the relationship between
the taste components and the taste was not thoroughly correlated and
investigated in most of these studies.

Detailed research on the relationship between the taste
components and the taste of sea urchin, shrimp and crab led to the
identification of the characteristic taste components. The characteristic components of sea urchin are Gly, Ala, Val, Glu, Met,
inosine 5'-monophosphate (IMP) and guanosine 5'-monophosphate (GMP)
(51). The contribution of Gly and Ala to sweetness, Val to bitter
ness, and Glu, IMP and GMP to umami taste was found. Met was shown
to be responsible for the characteristic taste of sea urchin. The
characteristic taste of shrimp is sweet taste which is attributed
to Gly, the largest component of all the free amino acids in shrimp
(52). The 12 components - Gly, Ala, Arg, Glu, CMP, AMP, GMP, Na^+ ,
K^+, Cl^-, PO_4^{3-} and betain - were shown to contribute to the
characteristic taste of boiled crab extract (53).

Free amino acids and peptides released by such proteolytic
enzymes as chymosin and lactic acid bacterial proteases in cheeses
contribute to the formation of cheese taste. Biede and Hammond (54)
reported that free amino acids and small peptides played an important

Table 9. Contents of Free Amino Acids in Sea Foods (mg/100 g)

Amino acid	Plaice	Yellowfin tuna	Cuttle-fish	Abalone	Scallop	Prawn	Snow crab
Tau	171	26	160	946	176	150	243
Asp	+	1	+	9	+	+	10
Thr	4	3	9	82	38	13	14
Ser	3	2	27	95	6	133	14
Gln + Asn	1	–	–	–	–	–	+
Glu	6	3	3	109	99	34	19
Pro	1	2	749	83	36	203	327
Gly	5	3	832	174	613	1222	623
Ala	13	7	181	98	82	43	187
Cys	–	–	3	–	3	+	–
Val	1	7	3	37	10	17	30
Met	1	3	7	13	12	12	19
Ile	1	3	6	18	3	9	29
Leu	1	7	12	24	0.3	13	30
Tyr	1	2	8	57	2	20	19
Phe	1	2	2	26	4	7	17
Trp	–	–	5	20	–	+	10
His	1	1220	16	23	10	16	8
Lys	17	35	15	76	7	52	25
Arg	3	0.6	246	299	935	902	579

+, trace; –, not detected.

role in producing the sweet and brothy tastes of Swiss cheese, and that medium sized (tri to hexa) peptides played an important role in bitterness. A number of studies on bitter peptides of cheese have been carried out. Several bitter peptides were isolated from different cheeses and their structures were determined. L-Leu-Trp-OH, a bitter peptide, was isolated from Swiss cheese (55), and Pro-Phe-Pro-Gly-Pro-Ile-Pro-Asn-Ser from Butterkäse (56). From Cheddar cheese were isolated Pro-Phe-Pro-Gly-Ile-Pro, Pro-Phe-Pro-Gly-Pro-Ile-Asn-Ser, and Gln-Asp-Lys-Ile-His-Pro-Phe-Ala-Gln-Thr-Gln-Ser-Leu-Val-Tyr-Pro-Phe-Pro-Gly-Pro-Ile-Pro (57). Recently, the formation mechanisms of free amino acids and peptides contributing to cheese taste have been shown (58). As shown in Fig. 2., the peptide, αSI-CN(f1-23), obtained from Gouda-type cheese was produced by the action of chymosin with αSI-casein(-CN). This peptide was degraded by lactic acid bacterial protease and small peptides, including three major peptides, αSI-CN(f1-9), αSI-CN(f1-13) and αSI-CN(f1-14), were formed. These small peptides were further degraded into smaller peptides and free amino acids by aminopeptidase of lactic acid bacteria.

It is said that flavor of beef, pork and chicken is improved by storage at a low temperature for given periods. We examined the effect of the storage at low temperature on the taste of meats (59). After beef, pork and chicken were stored at 4°C for 8, 5 and 2 days, respectively, the changes in intensity and levels of brothy taste and

Peptide content (% of total N)
15
10
5
0 2 5 10 20 35 50
Storage time (days)

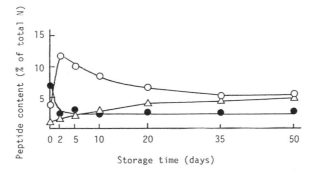

Fig. 1. Changes in various peptides during the storage of miso. A.P.L., the average number of amino acid residues in the peptides.
○—○, A.P.L. 3-4; ●—●, A.P.L. 4-6; △—△, A.P.L. 13-20.

αS1-casein(-CN) and αS1-CN(199)
chymosin
αS1-CN(1-23) + αS1-CN(24-199)
proteinases (lactic acid bacteria)
Peptides + αS1-CN(1-9) + αS1-CN(1-13) + αS1-CN(1-14)
aminopeptidases (lactic acid bacteria)
Free amino acids + Oligopeptides

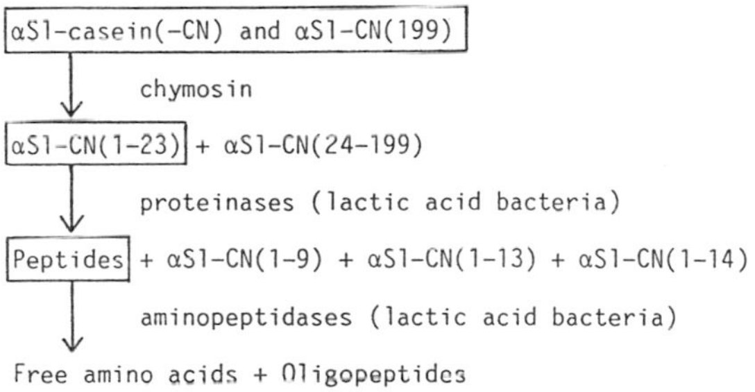

Fig. 2. Pathway of αSI-casein degradation during Gouda-type cheese ripening.
▭, protein or peptides subjected to hydrolysis.

taste components were examined. Examination of the brothy taste
intensity of meat before and after storage showed that the brothy
taste intensity of pork and chicken was significantly stronger after
storage than before. There was no significant difference in the
brothy taste intensity before and after storage of beef (Table 10).

Table 10. Effect of Additional Storage on the Intensity of the
Brothy Taste of Beef, Pork and Chicken

Meat	The no. of samples judged to have a more intense brothy taste :		n	Difference[a]
	Before additional storage	After additional storage		
Beef	12	4	16	NS
Pork	2	14	16	*
Chicken	8	23	31	*

a: NS, not significant; *, significant ($p < 0.05$).
SOURCE: Reprinted with permission from ref. 60. Copyright 1988 Marcel
Dekker.
The changes of various taste components of meats during storage were
also determined. In pork and chicken, especially, the differences in
the levels of free amino acids and peptides before and after storage
were very large (Fig. 3, 4), and these results corresponded to the
higher brothy taste intensity in pork and chicken after storage of
meat as compared to before storage. This correspondence suggests
that the increase in free amino acids and peptides contributed to the
improvement of meat taste after storage.
     Sensory evaluation of the relative strengths of each taste
(sweet, sour, umami, salty, bitter and brothy) among beef, pork and
chicken soups prepared after storage showed that the intensity of
umami and brothy tastes was weakest in beef soup (Fig. 5) (Rhue,
M.R., University of Tokyo, unpublished data.). There was less
Glu in beef than in pork and chicken. The addition of Glu into beef
soup to bring up the Glu concentration equal to those in pork and
chicken soups made the umami and brothy tastes in the beef soup
similar to those in pork and chicken soups. From this observation,
Glu seems to play a very important role in the umami and brothy
tastes of meats. This experiment showed that other free amino acids
also contribute somewhat to the meaty taste.
     The role of peptides in the taste of meats after storage is not
at all clear though peptides seem to contribute to the improvement of
the taste of stored meat as mentioned above. The role of Car and
Ans found in significant amounts in meats is also unknown. These
problems must be studied for clarification.
     Each food has a characteristic taste which is determined by the
balance of the primary and/or secondary tastes. Free amino acids and
peptides play an important role in the elicitation of each food
taste.

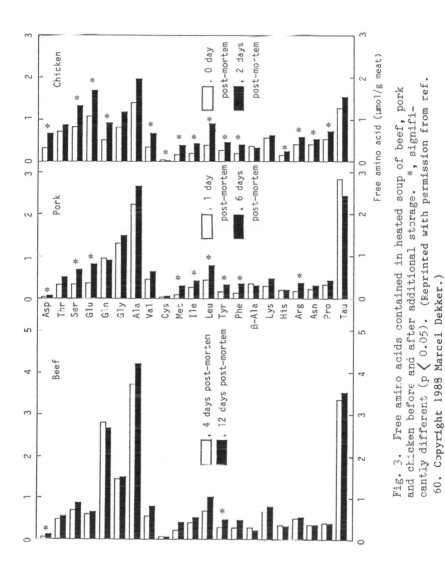

Fig. 3. Free amino acids contained in heated soup of beef, pork and chicken before and after additional storage. *, significantly different (p < 0.05). (Reprinted with permission from ref. 60. Copyright 1988 Marcel Dekker.)

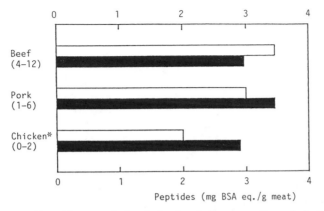

Fig. 4. Peptides contained in heated soup of meat before and
after additional storage.
The levels of peptides were obtained from the difference between
the values of phenol reagent-positive materials before and after
the addition of Cu^{2+} into the phenol reagent. ⬜, before
additional storage; ⬛, after additional storage;
*, significantly different ($p < 0.05$); numbers in parentheses,
time post-mortem (days). (Reprinted with permission from ref. 60.
Copyright 1988 Marcel Dekker.)

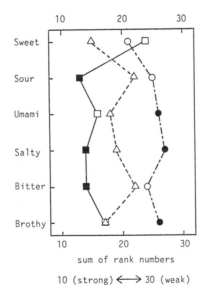

Fig. 5. Relative strength of each taste among beef, pork and
chicken soups.
The NaCl concentration of each soup was adjusted to 0.508%.
○--○, beef; △--△, pork; □—□, chicken. Closed symbols,
significantly different from others ($p < 0.05$). (Reprinted with
permission from ref. 60. Copyright 1988 Marcel Dekker.)

LITERATURE CITED

1. K. Ikeda, J. Tokyo Chem. Soc., 1908, 30, 820.
2. J. Solms, J. Agric. Food Chem., 1969, 17, 686.
3. J. Kirimura, A. Shimizu, A. Kimizuka, T. Ninomiya and N. Katsuya, J. Agric. Food Chem., 1969, 17, 689.
4. T. Takemoto and T. Nakajima, J. Pharm. Soc. Japan, 1964, 84, 1230.
5. T. Takemoto, T. Nakajima and T. Yokobe, ibid., 1964, 84, 1232.
6. S. Yamaguchi, J. Food Sci., 1967, 32, 473.
7. T. Yokotsuka, N. Saito, A. Okuhara and T. Tanaka, Nippon Nogeikagaku Kaishi, 1969, 43, 165.
8. T. Tanaka, N. Saito, A. Okuhara and T. Yokotsuka, Nippon Nogeikagaku Kaishi, 1969, 43, 171.
9. R.H. Mazur, J.M. Schlatter and A.H. Goldkamp, J. Amer. Chem. Soc., 1969, 91, 2684.
10. H. van del Wel and K. Loeve, Eur. J. Biochem., 1985, 31, 221.
11. J.A. Morris and R.H. Cagan, Biochim. Biophys. Acta, 1972, 261, 114.
12. K.H. Ney, Z. Lebensm.-Unters. Forsch., 1971, 147, 64.
13. T. Matoba and T. Hata, Agric. Biol. Chem., 1972, 36, 1423.
14. M. Fujimaki, M. Yamashita, Y. Okazawa and S. Arai, Agric. Biol. Chem., 1968, 32, 794.
15. M. Yamashita, S. Arai and M. Fujimaki, Agric. Biol. Chem., 1969, 33, 321.
16. M. Fujimaki, M. Yamashita, Y. Okazawa and S. Arai, J. Food Sci., 1970, 35, 215.
17. S. Arai, M. Yamashita, H. Kato and M. Fujimaki, Agric. Biol. Chem., 1979, 34, 729.
18. H. Wisser and H.D. Belitz, Z. Lebensm.-Unters. Forsch., 1975, 159, 329.
19. K.M. Clegg, C.L. Lim and W. Manson, J. Dairy Res., 1974, 41, 283.
20. T. Matoba, R. Hayashi and T. Hata, Agric. Biol. Chem., 1970, 34, 1235.
21. N. Minamiura, Y. Matsumura, J. Fukumoto and T. Yamamoto, Agric. Biol. Chem., 1972, 36, 588.
22. N. Minamiura, Y. Matsumura and T. Yamamoto, J. Biochem., 1972, 72, 841.
23. H. Shiraishi, K. Okuda, Y. Sato. N. Yamaoka and K. Tuziura, Agric. Biol. Chem., 1973, 37, 2427.
24. K. Otagiri, T. Shigenaga, H. Kanehisa and H. Okai, Bull. Chem. Soc. Jpn., 1984, 57, 90.
25. I. Miyake, K. Kouge, H. Kanehisa and H. Okai, Bull. Chem. Soc. Jpn., 1983, 56, 1678.
26. K. Otagiri, I. Miyake, N. Ishibashi, H. Fukui, H. Kanehisa and H. Okai, Bull. Chem. Soc. Jpn., 1983, 56, 1116.
27. M. Tada, I. Shinoda and H. Okai, J. Agric. Food Chem., 1984, 32, 992.
28. T. Huynh-ba and G. Philippossian, J. Agric. Food Chem., 1987, 35, 165.
29. S. Arai, M. Yamashita, M. Noguchi and M. Fujimaki, Agric. Biol. Chem., 1973, 37, 151.
30. S. Ohyama, N. Ishibashi, M. Tamura, H. Nishizaki and H. Okai, Agric. Biol. Chem., 1988, 52, 871.

31. M. Fujimaki, S. Arai, M. Yamashita, H. Kato and M. Noguchi,
     Agric. Biol. Chem., 1973, 37, 2891.
32. M. Noguchi, S. Arai, M. Yamashita, H. Kato and M. Fujimaki,
     J. Agric. Food Chem., 1975, 23, 49.
33. Y. Yamasaki and K. Maekawa, Agric. Biol. Chem., 1978, 42, 1761.
34. Y. Yamasaki and K. Maekawa, Agric. Biol. Chem., 1980, 44, 93.
35. M. Suyama and T. Shimizu, Bull. Japan. Soc. Sci. Fish., 1982,
     48, 89.
36. K. Oishi, Bull. Japan. Soc. Sci. Fish., 1969, 35, 232.
37. M. Noguchi, M. Yamashita, S. Arai and M. Fujimaki, J. Food Sci.
     1975, 40, 367.
38. A. Matsushita and A. Yamada, Nippon Nogeikagaku Kaishi, 1957,
     31, 578.
39. Z. Nagashima, M. Nakagawa, H. Tokumaru and Y. Toriumi, Nippon
     Nogeikagaku Kaishi, 1957, 31, 169.
40. Y. Sakado, Nippon Nogeikagaku Kaishi, 1950, 23, 262.
41. Y. Sakado and F. Hashizume, Nippon Nogeikagaku Kaishi, 1950, 23,
     269.
42. T. Take and H. Otsuka, J. Jap. Soc. Food Nutri., 1967, 20, 169.
43. R. Buri, V. Signer and J. Solms, Lebensm.-Wiss. Technol., 1970,
     3, 63.
44. S. Yamaguchi, In Umami: A Basic Taste, Y. Kawamura and M.R.
     Kare (Ed.) Marcel Dekker, New York, 1987, p 41.
45. H. Ito and H. Ebine, Norinsyo Syokuryo Kenkyusho Hokoku, 1965,
     19, 121.
46. S. Hondo, I. Ouchi and T. Mochizuki, Nippon Shokuhinkogyogaku
     Kaishi, 1969, 16, 155.
47. Y. Yamasaki, Nippon Kaseigaku Kaishi, 1987, 38, 93.
48. K. Takahashi, M. Tadanuma, K. Kitamoto and S. Sato, Agric.
     Biol. Chem., 1974, 38, 927.
49. S. Konosu, in Shinsuisan Handbook, R. Kawashima, S. Tanaka,
     H. Tsukahara, M. Nomura, M. Toyomizu and Y. Asada(Ed.),
     Kodansha, Tokyo, 1981, p 465.
50. Y. Komata, N. Kosugi and T. Ito, Bull. Japan. Soc. Sci. Fish.,
     1962, 28, 623.
51. Y. Komata, Bull. Japan. Soc. Sci. Fish., 1964, 30, 749.
52. M. Fujita, The Doctoral Thesis, 1961, p 34.
53. T. Hayashi, K. Yamaguchi and S. Konosu, J. Food Sci., 1981,
     46, 479.
54. S.L. Biede and E.G. Hammond, J. Dairy Sci., 1979, 62, 238.
55. Y. Guigoz and J. Solms, Lebensmitt.-Wiss. Technol., 1974, 7,
     356.
56. L. Huber and H. Klostermeyer, Milchwissenschaft, 1974, 29, 449.
57. J.S. Hamilton and R.D. Hill, Agric. Biol. Chem., 1974, 38, 375.
58. S. Kaminogawa, T.R. Yan, N. Azuma and K. Yamauchi, J. Food Sci.,
     1986, 51, 1253.
59. T. Nishimura, M.R. Rhue and H. Kato, Agric. Biol. Chem., 1988,
     52, 2323.
60. Nishimura, T. and Kato, H. Food Reviews International, 1988,
     4(2), 175-194.

RECEIVED October 6, 1988

RECENT INVESTIGATIONS
OF SPECIFIC FLAVORS

Chapter 14

New Dimensions in Flavor Research

Herbs and Spices

Braja D. Mookherjee, Richard A. Wilson, Robert W. Trenkle, Michael J. Zampino, and Keith P. Sands

International Flavors and Fragrances, Research and Development, 1515 Highway 36, Union Beach, NJ 07735

Herbs and spices are not only common household food ingredients but also integral parts of various flavor and fragrance creations. Most of the spices used day-to-day are either dried or aged, but few people are aware of the fact that live spices have different aroma properties from those that are dead. The chemical differences in the aroma profiles of "living" vs dead leafy spices have now been characterized. The methodology and analytical results for some common spices are presented.

Today, in western society, we take herbs and spices for granted, but this was not true in the past where they were rare and prized commodities. Herbs and spices were so precious that even a slave could be bought for a handful of spice (1).

Generally speaking, the important spices came from the East, especially from India, Ceylon, and the eastern Spice Islands. Beginning with Marco Polo, various travelers like Vasco da Gama ventured eastward, found the lands of spice, and opened the door to the West for the spice trade. England eventually became the center for the European spice trade. It should be mentioned in this connection that, in the late 17th Century, the Americans also benefited from the spice trade. Boston-born Elihu Yale went to England where he worked in the British East India Company which held a monopoly on all trade with India and whose ships brought the first cargo of cinnamon. He eventually became Governor of Madras, India and acquired a fortune which he donated to a university in Connecticut which now bears his name and is known as Yale University (1).

From ancient times up until modern days herbs and spices have played a dynamic role in our daily lives. When we clean our teeth in the early morning with toothpaste we encounter mint oils. When we wash our bodies and clothes with soaps and detergents we find the essences of rosemary and lavender. At midday and in the evening on the dinner table the smells of spices elevate the appetite. More people than ever are discovering the secrets of great cuisine which rely heavily on herbs and spices. In addition

0097–6156/89/0388–0176$06.00/0
© 1989 American Chemical Society

to the use of herbs and spices in the culinary arts, the back-to-nature health movement has also called our attention to these materials. People are buying more and more herb and spice products, however, they are not aware of the fact that most of these products have been processed and the aromas are entirely different from those found in the living plant. The reason is that when the umbilical cord connecting the fruit, flower, leaf, or seed to the plant is severed, these products are then essentially dead and their aroma is perceptibly changed.

The chemical differences in the aroma profile between living and dead fruits, flowers, herbs, and spices have now been characterized, and the methodology and analytical results for several common examples will now be described. The first example will deal with typical results obtained for a living vs dead flower. The first flower chosen not only has the highest priority in the fragrance industry but also is heavily used for flavoring foods by Oriental people. This flower is none other than the rose which was called the "Queen of Flowers" by the Greek poetess, Sapho, in 600 B.C. The rose most likely originated in China and was introduced into Spain from China by invading Arabs in the 7th Century and into India in the 10th Century (2). The rose is prized chiefly for its blossoms, and, though it has been known since ancient times for the making of fragrance, it was Empress Nurjuhan, the wife of Indian Emperor Jahangir, in the 13th Century who first made attar of rose by spreading rose petals on her morning bath water.

Of the 200 varieties of rose the most coveted for the making of Otto of Rose for fragrance use is <u>Rosa-damascena</u> which comes from Bulgaria. It takes about 4000 pounds of rose to produce one pound of Rose Otto; hence the cost of $2500 per pound. Although most roses grown for commercial oil production come from Bulgaria or the south of France, in the 1930's American horticulturists started to breed hybrid tea roses for both their form and fragrance. Many of them have unique aromas in their own right, although none is the equal of <u>Rosa damascena</u>. One of the best from the point of view of aroma, yellow tea rose (J.F.K.), was chosen for analysis.

Two side-by-side experiments were performed on the yellow tea rose; first on the picked blossoms and next on the blossoms still attached to the plant. The latter is called the "living flower" analysis. (Duplicate experiments were performed on other blossoms from the same plant and on blossoms from other plants of the same species and variety. In all cases, there were no significant differences observed in the analytical data.)

In the method of analysis routinely employed on picked flowers, the blossoms are placed in a flask equipped with a trap packed with Tenax GC. The flask is purged with air for 6-12 hours depending on the type of flower, and the volatiles are collected on the Tenax and then desorbed into the gas-liquid chromatograph for analysis by GC/MS. In the method of analysis used for the living flower, one single living blossom is placed into a suitable glass chamber which contains a Tenax trap on one sidearm. Air is drawn over the blossom and through the Tenax trap by a pump under the same conditions as employed for the picked flower. In this way, the aroma profiles of many different flowers were compared.

The comparative headspace analysis of living vs picked yellow tea
rose (J.F.K.) is shown in Table I.

Table I. Major Differences Between Living and
Picked Yellow Tea Rose (J.F.K.) Flower

Compound	Living Rose Flower % (AN)	Picked Rose Flower Air Purged % (AN)
cis-3-Hexenyl acetate	20.7	5.4
Hexyl acetate	8.4	4.3
Phenyl ethyl alcohol	5.7	3.3
Phenyl ethyl acetate	5.5	1.5
3,5-Dimethoxy toluene	10.0	18.6
alpha Elemene	-	4.1
Geranyl acetone	2.2	-
Dihydro beta ionol	-	2.6
alpha Caryophyllene	0.3	2.1
alpha Farnesene	5.8	3.0

These data reveal that the composition of the picked tea rose
is very much changed from that of the living rose. As one can
see, cis-3-hexenyl acetate which constitutes 21% of the living
rose headspace volatiles is drastically reduced to 5% in the
picked rose. At the same time, 3,5-dimethoxy toluene, one of the
character-donating components of tea rose, is dramatically doubled
in the picked flower, whereas important constituents like phenyl
ethyl alcohol and its acetate are reduced in the picked flower.

In this way, many other common and uncommon flowers such as
jasmine, narcissus, osmanthus, honeysuckle, hyacinth, lily-of-the-
valley, lilac, and tuberose have been analyzed. In all cases,
considerable differences have been observed in the aroma profiles
of the living and picked flowers. In connection with the living
flower analytical program, the concept was also extended to the
flavor field, especially fruits. The fruits and flowers work has
served as the basis of a recent report to the 10th International
Congress of Essential Oils in Washington, D.C. in 1986.

The first subject to be tested in the fruits area was peach
due to the importance of its flavor and aroma to both the flavor
and fragrance industry. The peach actually originated in China
though botanists thought that it came from Persia, hence its name,
Prunus persica (Persian Plum-Tree). From China, its cultivation
spread west to Persia in the 3rd Century B.C. eventually reaching
Europe. From there, the Spanish introduced the peach to the New
World where the American Indians developed a taste for the fruit,
even naming one of their thirteen months for it. Thomas
Jefferson, a peach lover, planted peach trees at his birthplace,
Monticello, in Virginia, when he became President (3). This is a
brief history of the peach, and now the analysis of the living
peach will be described.

A peach still attached to the tree was selected for analysis
on the basis of its possessing a full, rich, at-the-peak-of-
ripeness aroma. Taking care not to bruise the fruit, the peach

was placed into a flask which was designed to handle larger objects. With the Tenax trap and pump in place, the respiration gases of the fruit were collected for 16 hours. A peach of equal ripeness was harvested from the same tree and immediately set up for collection of its volatiles. The major differences between the headspace volatiles of living and picked peach are shown in Table II.

Table II. Major Differences Between Living and Picked Peach

Compound	Living Peach % (AN)	Picked Peach Air Purged % (AN)
Ethyl acetate (a)	6.2	–
Dimethyl disulfide	0.6	–
cis-3-Hexenyl acetate (a)	9.7	–
Methyl octanoate	34.2	7.1
Ethyl octanoate	7.4	11.0
6-Pentyl alpha pyrone	trace	10.6
gamma Decalactone	2.5	39.2

(a) Identified for the first time in peach

One can observe that the major volatiles of living peach are lower boiling with methyl octanoate, now identified for the first time in peach, predominating. The identification of dimethyl disulfide for the first time in peach is of interest. Very little peach lactone and pentyl pyrone are seen in the living peach, whereas they are major components of the picked fruit. Methyl octanoate is considerably decreased and the lower boiling constituents are essentially gone after picking.

It is reasonable to expect that what is true for living and picked fruits and flowers could also be true for herbs and spices, although it is possible to keep herbs and spices in an acceptable olfactory condition for longer periods of time than one can preserve picked fruits and flowers. The first subject for testing of this theory was mint because of its extensive use in the flavor industry. American spearmint, Mentha spicata, will be described first. It is interesting to note that the word "mint" was coined by the early Greeks after the mythical character MINTHE. The term "mint" refers to the dried leaf of the spearmint plant, which, available in flake or extract form, has an aromatic, sweet flavor with cool aftertaste (4).

Interestingly, American spearmint is not native to North America but was introduced from Europe during the 17th Century and has since been widely grown. Millions of pounds of this oil have been produced in this country due to its extensive and popular use as a flavoring ingredient, particularly in chewing gums and toothpastes.

The technique of headspace analysis of the living and picked American spearmint plant is the same as in the case of living and picked flowers and fruits. The picked spearmint was taken from the same plant used for the living plant analysis. In order to

simulate the commercial process for making spearmint oil, freshly
picked stems and leaves were kept at room temperature for 24 hours
with a weight loss of 50%.   This semi-dried material was then
analyzed for headspace volatiles and compared with that of living
plants.   Table III represents the comparative analysis of living
vs picked spearmint plant and, for purposes of comparison, a
typical commercial oil.

Table III. Major Differences in Spearmint Volatiles

Commercial	Living Plant % (AN)	Picked Plant Air Purged % (AN)	Comm. Oil
Hexanal	0.5	trace	–
Hexanol	–	2.3	0.1
Limonene	17.7	1.8	21.4
Dihydro carvone	0.7	2.6	0.1
Carvone	24.0	70.0	63.0
Menthone/isomenthone	–	–	1.2
Menthol isomers	–	–	1.7
1,3,5-Undecatriene (a) (mixture of 4 isomers)	0.5	–	–

(a)   Identified for the first time in spearmint volatiles

        Interestingly,   neither   the   isomeric   menthones   nor   the
isomeric menthols were detected in the living or picked plant
material, but they are both present in appreciable amounts in
commercial oil.   At the same time, a very powerful green odorous
compound, 1,3,5-undecatriene (isomer mix), has now been identified
for the first time in the living spearmint to the extent of 0.5%.
One can also observe that carvone, the true character-donating
component of spearmint oil, constitutes 70% of the total headspace
volatiles of picked spearmint but only 24% of those of the living
plant.   The opposite is true in the case of limonene which is only
a minor constituent in the picked plant but a major component of
the   living   mint.   These   variations   in   constituents   will
drastically influence the odor of living spearmint.
        After spearmint, naturally comes peppermint, Mentha piperita,
as distinguished from the many other species of Mentha herb
including Mentha spicata.   This herb is native to Europe and has
become naturalized in North America.   Of the many hybrids of
peppermint, only two varieties, black and white, are commonly
grown.   Of these, black peppermint, also known as English
peppermint, is the variety most extensively grown in the United
States because of its hardiness and high oil yield.   The United
States is the world's largest peppermint oil producer, and the oil
is mainly and extensively used for oral hygiene products, chewing
gum and confectioneries.   Pure peppermint oil has a very agreeable
odor and a powerful, aromatic taste followed by a sensation of
cold when air is drawn into the mouth (5).   Table IV represents
the comparative analysis of living vs picked peppermint plant and
commercial   oil.   These   experiments   were   performed   on   black
peppermint.

Table IV.   Comparative Analysis of Peppermint Volatiles

Compound		Living Plant % (AN)	Picked Plant Air Purged % (AN)	Comm. Oil
Hexanal	(a)	–	0.1	–
cis-3-Hexenal		–	0.5	–
trans-2-Hexenal		–	0.8	–
cis-3-Hexenol	(a)	–	0.3	–
trans-2-Hexenol	(a)	–	1.4	–
Hexanol	(a)	–	0.5	–
2,4-Hexadienal	(a)	–	0.1	–
1-Octen-3-ol	(a)	–	2.0	–
Eucalyptol		–	–	5.7
Menthone		0.2	12.7	18.1
isoMenthone		9.6	7.7	2.3
Menthofuran		49.7	26.3	5.2
neoMenthol		–		1.7
Menthol		trace	4.7	44.2
neoisoMenthol		–	–	1.9
isoMenthol		–	–	0.2
Pulegone		1.6	24.5	1.7
1,3,5-Undecatriene (mixture of 4 isomers)	(a)	0.6	–	

(a)   Identified for the first time in peppermint volatiles

As one could see, tremendous differences exist between the living and picked peppermint volatiles. For example, six-carbon alcohols and aldehydes are present only in the picked plant. Interestingly, these very green compounds have not been identified before in peppermint. At the same time, 1-octen-3-ol, which possesses an earthy, mushroom odor and which has not been previously found in peppermint, was identified only in the picked herb. In the class of compounds responsible for the cooling effect of peppermint oil, surprisingly, menthone is a major component in the picked herb but is only present in trace quantities in the living plant, whereas isomenthone, the more powerful of the two isomers, is present to approximately the same extent in both. Menthofuran, which has the reputation of being a less desirable component of mint oils, surprisingly constitutes 50% of the volatiles of the living plant decreasing to 26% in the picked plant material. In the opinion of the present authors, menthofuran imparts a characteristic fresh mintiness. On the other hand, menthol, the classical cooling compound, is present in negligible amounts in the living plant while its content varies from 5-45% in picked and commercial oils respectively. Pulegone, another characteristic component with a sweet, weedy, minty odor, is a major component only in the picked peppermint. Finally, as with spearmint, the powerful, diffusive herbaceous – green compound, 1,3,5-undecatriene (isomer mix), was detected only in the living peppermint and at a relatively high level (0.6%) considering its strength. This compound has never before been identified in peppermint. Now, one can easily see that a living

peppermint leaf has quite a different aroma from that of dried
peppermint or commercial peppermint oil.

The next herb which was analyzed was rosemary, Rosmarinus
officinalis, not because the herb finds extensive use as a
seasoning, condiment, or meat flavor, which it does, but because
it is widely employed in fragrancing colognes, toilet waters and
household products such as soaps and detergents. Rosemary is rich
in legend and tradition. It is said to have been used as early as
5000 B.C. One legend claims that rosemary will grow only in the
gardens of the righteous (6). Even in Shakespeare's "Hamlet",
Ophelia said, "There is rosemary; that's for remembrance." In
England to this day rosemary is placed on the graves of heroes so
that their memories will be eternal. Its name is derived from
"ros maris" which means "sea-dew" (7), and it indeed grows near
the sea in Spain, Dalmatia, Morocco, Tunisia, and Turkey.

Table V shows the comparative analysis of the headspace of
living and picked rosemary and a typical commercial oil.

Table V. Comparative Analysis of the Volatiles of Rosemary

Compound		Living Plant % (AN)	Picked Plant Air Purged % (AN)	Comm. Oil
trans-2-Hexenal	(a)	–	0.2	–
cis-3-Hexenol	(a)	–	0.7	–
Hexanol	(a)	–	0.3	–
alpha Pinene		1.1	0.7	13.3
Myrcene		9.5	11.1	1.7
beta Pinene		0.2	0.1	7.6
para Cymene		19.8	13.7	1.7
Limonene		14.1	14.3	1.0
Eucalyptol		2.0	0.7	44.5
Linalool		7.1	7.6	0.8
Camphor		0.2	–	10.1
Estragole	(a)	3.0	0.3	trace
cis-Carveol	(a)	0.2	0.6	–
Citronellol	(a)	0.6	1.1	–
alpha Campholenic alcohol and acetate	(a)	0.3	0.8	–

(a) Identified for the first time in rosemary volatiles

Again, one could easily see the qualitative and quantitative
differences between the living and picked rosemary. It is
interesting to observe that the very fatty-green components,
trans-2-hexenal, cis-3-hexenol, and hexanol, are present only in
the picked plant and were not detected in the living herb. At the
same time, hydrocarbon constituents, alpha pinene, myrcene, beta
pinene, para cymene, and limonene, do not vary much from living to
picked. However, the most interesting observation from our
experiment is that eucalyptol and camphor, which traditionally are
the major constituents of commercial rosemary oil as shown here,
(45 and 10% respectively), are present in both living and picked
rosemary only in very small quantities, 0.7-2% for eucalyptol and

0.2% for camphor which was not found in the picked plant at all. The present authors are, therefore, of the opinion that eucalyptol and camphor do not play major roles in producing the fresh rosemary odor whereas estragole, a newly reported rosemary constituent possessing a powerful sweet, herbaceous aroma, probably is in large part responsible for the fresh herbaceous rosemary character. It is also of interest to note the identification for the first time in rosemary of cis-carveol, citronellol, and alpha campholenic alcohol and its acetate and to observe that the quantities of each increase on picking.

The next herb chosen for analysis was thyme, Thymus vulgaris, which is also native to southern Europe and the Mediterranean and is cultivated in the southern United States as well. Thyme is used extensively in flavors for food products such as sauces, dressings, pickles, and canned meats as well as in pharmaceutical preparations. The excellent germicidal properties of the phenols of the oil are exploited in oral hygiene products such as gargles and mouthwashes and in numerous disinfectants. Cough syrups and lozenges are often activated with thyme oil. In perfumery, it finds use in soaps and detergents for its freshness with hints of medicinal notes. Table VI shows the comparative analysis of the headspace volatiles of thyme.

Table VI. Comparative Analysis of the Volatiles of Thymus vulgaris

Compound	Living Plant % (AN)	Picked Plant Air Purged % (AN)	Comm. Oil
trans-2-Hexenal [a]	–	2.8	–
?,4-Hexadienal [a]	–	0.1	–
cis-3-Hexenyl acetate [a]	11.2	0.1	–
1-Octen-3-ol	8.2	8.0	–
para Cymene	30.0	50.0	30.0
Limonene	1.3	1.1	1.7
Thymol methyl ether	–	1.3	–
Carvacrol methyl ether	–	1.5	0.1
Thymol	15.2	9.0	39.7
Carvacrol	1.5	0.9	1.0

[a] Identified for the first time in thyme volatiles

As has been observed in the case of peppermint and rosemary, fatty-green six-carbon components like trans-2-hexenal and 2,4-hexadienal are again found only in the picked thyme and are not detected at all in the living plant. It may be concluded that these components are actually formed by enzymatic oxidation during the overnight drying process. The fresh herbaceous quality of the living plant is probably not associated with these components, but is, at least in part, due to the true green aroma of compounds like cis-3-hexenyl acetate which occurs to the extent of 11% in the living plant but is only a trace component of the picked herb. At the same time, the characteristic aroma components of thyme oil, thymol and carvacrol, are both present to a greater extent in

the living herb. Interestingly, the corresponding methyl ethers
are detected only after picking. However, para-cymene increases
from 30-50% on picking.

Evergreen Cassia, Cinnamomum cassia also known as Chinese
Cinnamon, is native to China, Burma, and many sub-tropical
countries. It has a long, shiny leaf, a small pale green flower,
and a loose peeling bark. The trees are grown in plantations and
are coppiced for the new long shoots which provide the scented
bark. The stripped bark curls into quills as it dries and is
exported in bundles. The dried unripe fruits are sold as Chinese
cassia buds, and the dried leaves and stems are used to distill
cassia oil. Totally dried leaves are used as a flavoring "Tej
Pat" in the day-to-day Indian cookery. Cassia has been used as a
spice in Europe since the Middle Ages and it has also found use in
the treatment of indigestion and to increase the flow of mother's
milk (8). We chose to compare the volatiles of fresh leaves and
aged leaves dried to 50% weight along with commercial oil. The
results for cassia leaves freshly harvested in Hawaii are shown in
Table VII.

Table VII. Comparative Analysis of Volatiles of Cassia Leaf

Compound	Fresh Leaf % (AN)	Aged Leaf Air Purged % (AN)	Comm. Oil
trans-2-Hexenal [a]	4.0	0.8	–
Phenyl ethyl alcohol	2.1	0.1	0.4
trans-Cinnamaldehyde	50.0	70.0	70.0
Cinnamyl alcohol	20.6	0.3	0.3
Coumarin	7.9	4.4	1.7
2-Methoxy cinnamaldehyde	1.0	1.4	11.5
4-Methoxy cinnamaldehyde [a]	4.1	12.3	–

[a]   Identified for the first time in cassia volatiles

Table VIII shows the dramatic differences between fresh and
dried leaves. In this case, as opposed to peppermint, rosemary
and thyme, trans-2-hexenal is more in the fresh than in the dried
and aged. The same is true for phenyl ethyl alcohol.
Interestingly, trans-cinnamic aldehyde constitutes 50% of the
total living headspace volatiles, but it is still less than in the
aged leaf and commercial oil. However, cinnamyl alcohol
represents 20% of the fresh volatiles but is only a trace
component of the aged leaf and oil. 4-Methoxy cinnamic aldehyde,
identified as a cassia constituent for the first time, also
increases 3-fold on drying but has disappeared completely in the
commercial oil. 2-Methoxy cinnamic aldehyde, sometimes called the
character impact component of cassia oil, is present in the
headspace of the leaves to only a minor extent but it is the
second most abundant component of the oil.

An herbal plant, a part of which is also considered as a
spice, is the last subject to be described. This plant is
coriander, Coriandrum sativum. Even though the name, "coriander"
originates from the Greek "koriannon" meaning "bug", a reference

to the smell of the leaves and unripe seed, actually it has been cultivated for thousands of years in India, China and Egypt (8). Even today the seeds and leaves are used daily in Oriental countries for flavoring cooking. But, at the same time, oil derived from dried seeds is an important ingredient in modern perfumery, particularly in fine fragrances such as "Drakkar Noir".

Since, in the opinion of the present authors and contrary to the observations of the ancient Greeks, green coriander leaves and seeds have a very aesthetic aroma, the odor profiles of both living and dead leaves and seeds were compared. The data for coriander leaf volatiles are shown in Table VIII.

Table VIII. Comparative Analysis of Volatiles of Coriander Leaves

Compound		Living Leaf % (AN)	Picked Leaf Air Purged % (AN)	Comm. Herb Oil
Hexanal	(a)	0.4	2.0	trace
trans-2-Hexenal	(a)	0.3	3.2	–
Hexanol	(a)	–	1.1	trace
Nonane	(a)	15.2	4.7	3.6
Decanal		11.4	4.7	14.8
trans-2-Decenal	(a)	35.5	39.2	26.8
trans-2-Decenol	(a)	2.6	–	2.4
Decanol	(a)	2.5	–	1.3
Undecanal		1.5	–	0.7
2,4-Decadienal	(a)	0.1	–	–
Dodecanal		1.2	–	0.5
trans-2-Dodecenal	(a)	9.7	4.3	2.7
trans-2-Dodecenol	(a)	0.4	–	trace
Tetradecanal		0.1	–	trace
trans-2-Tetradecenal	(a)	3.7	4.2	trace

(a) Identified for the first time in coriander leaf volatiles

As in the case of peppermint, rosemary, and thyme, the content of six-carbon aldehydes and alcohols increases on picking and drying. Interestingly, a very common hydrocarbon, nonane, has been found for the first time in coriander in a high concentration (15%) in the living leaves. This compound is drastically reduced on picking. Surprisingly, nonane possesses a very characteristic fresh coriander leaf odor. The decanal content also decreases on picking and aging. Decanal is a very orangy chemical and, indeed, it is a character-donating component of orange oil and, thus, lends a citrus note to the living herb.

Contrary to all literature reports on coriander leaf volatiles, the major constituent reported here for the first time is trans-2-decenal, the content of which is relatively constant from living leaf to picked to commercial herb oil. Two other alpha, beta unsaturated aldehydes, trans-2-dodecenal and trans-2-tetradecenal, both reported also for the first time, are present in appreciable quantities in both the living and dried leaf and at lower levels in the oil. A large number of related aldehydes and alcohols, both saturated and unsaturated, many reported for the first time in coriander, were identified in the living leaf and in the oil but were not found on drying of the leaf.

The headspace volatiles of both living green coriander seed and picked, dried green seed were analyzed as well as commercial seed oil. Even though green coriander seed has quite a different odor from the oil obtained from ripe seed, these aroma profiles are presented in Table IX for the purpose of comparison.

Table IX. Comparative Analysis of the Volatiles of Coriander Seed

Compound	Living Green Seed % (AN)	Picked Dried Green Seed % (AN)	Comm. Seed Oil
alpha Pinene	0.1	0.3	5.1
Linalool	20.9	67.1	73.3
Camphor	0.3	–	4.8
3-Decenal (a)	0.4	–	–
Decanal (a)	3.1	0.1	–
trans-2-Decenal (a)	24.9	–	–
trans-2-Decenol (a)	3.8	–	–
trans-2-Undecenal (a)	3.7	–	–
trans-2-Dodecenal (a)	16.7	–	–
trans-2-Tetradecenal (a)	3.8	–	–
Geranyl acetate	10.0	12.5	2.2

(a)   Identified for the first time in coriander seed volatiles

As reported in the literature, linalool constitutes two-thirds of coriander seed oil volatiles. On the other hand, it is only 21% in the living green seed. At the same time, it is drastically increased to 67% on picking and drying of the seed. Geranyl acetate, which is one of the character impact components of coriander seed oil, is present to the extent of 10% in the living seed volatiles but is reduced to 2% in the commercial oil. It is obvious from the table that unsaturated aldehydes and alcohols are major constituents of the living green seed, but these compounds completely disappear after picking and also are absent in the commercial oil. In the opinion of the present authors, the unsaturated compounds shown in Table IX, which have not been reported before as constituents of coriander, are the character donating components of the green seed. They are, undoubtedly, justification for the Greek word "koriannon" for "bug".

It has been demonstrated and proved by ample examples from flowers, fruits, herbs and spices that the volatile constituents of living natural products differ considerably from those of the corresponding picked entities, justifying the assertion that the odor is completely different on picking. To our knowledge, this scientific observation has never been made before. With these data, in IFF we are creating new true-to-nature flavor and fragrance compositions.

Literature Cited

1. Collins, M. Spices of the World Cookbook by McCormick; Penguin Books: New York, N.Y., 1964; p 4.
2. Encyclopedia Britannica, 1968; Vol. 19, p 621.
3. Panati, C. Browsers Book of Beginnings; Houghton Miflin Co.: Boston, Mass., 1984; p 105.
4. Collins, M. Spices of the World Cookbook by McCormick; Penguin Books: New York, N.Y., 1964; p 34.
5. Encyclopedia Britannica, 1968; Vol. 17, p 587.
6. Collins, M. Spices of the World Cookbook by McCormick; Penguin Books: New York, N.Y., 1964; p 50.
7. Jesse, J. Perfume Album; Robert E. Krieger Publishing Co.: Huntington, N.Y., 1951; p 83.
8. Garland, S. The Herb and Spice Book; Francis Lincoln Publishers Limited: London, England, 1979; p 48.

RECEIVED October 25, 1988

Chapter 15

Flavor of Cooked Meats

Fereidoon Shahidi

**Department of Biochemistry, Memorial University of Newfoundland,
St. John's, Newfoundland A1B 3X9, Canada**

The spectrum of volatile flavor components of
cooked meats from different species was investigated.
The chemical nature of flavor volatiles was
representative of most classes of organic compounds.
Hexanal was found to be the predominant volatile
component in each case and its content was directly
proportional to the amount of TBA-reactive species,
while inversely proportional to the flavor
acceptability of meats. Nitrite curing depressed the
production of lipid oxidation products and nitrite-
free curing composition duplicated the action of
nitrite on meat, flavorwise.

Flavor is an important sensory aspect of the overall acceptability
of meat products. It is perceived as the simultaneous stimulation
of our taste and odor senses due to high molecular weight components
and volatile chemicals present in cooked meats. The overwhelming
effects of flavor volatiles has a tremendous effect on sensory
acceptability of foods even before they are consumed.

Meat from different species constitutes an integral part of our
diet (except for vegeterians) and provides us with a good source of
well-balanced amino acids. Although raw meat has little flavor and
only a blood-like taste, it is a rich reservoir of non-volatile
compounds with taste tactile properties, as well as flavor enhancers
and aroma precursors (1,2). Non-volatile precursors of cooked meat
flavor are water-soluble substances and these include amino acids,
peptides, reducing sugars and vitamins, particularly vitamin B_1
(thiamine). On cooking, free amino acids (e.g. cysteine/cystine)
are produced from the action of proteolytic enzymes which starts
during the post-mortem period; breakdown of glycogen results in the
production of glucose, fructose, etc. (3,4). These, or their
breakdown products, together with other low molecular weight water-
soluble matters such as thiamine or its breakdown products react
with one another. Often products of one reaction become precursors
for others. Thus, interactions of this type and specifically non-

0097–6156/89/0388–0188$06.00/0

enzymatic browning or Maillard reactions (5) lead to the formation
of a large number of important volatile chemicals which are
essential for cooked meat flavor development (6). These together
with a contribution from reactions of lipids play an important role
in the overall flavor of meat which is distinct from species to
species.

Fat portion of meats, particularly their phospholipid
components, undergo autoxidation/degradation (7) and produce an
overwhelming number of volatiles. Fats also serve as a depot of
fat-soluble compounds that volatilize on heating and strongly affect
flavor. Since compositional characteristics of lipids in meats,
vary from one species to another, these factors may be responsible
for the development of some species-specific flavor notes in cooked
meats (8,9). Obviously presence of 4-methyloctanoic and 4-
methylnonanoic acids with a mutton-specific flavor note is an
exception. These branched fatty acids are biosynthesized by sheep
(10,11). The swine sex odor compounds, mainly associated with
males, are another exception (12).

Presence and concentration of hemoproteins and free iron in
meats from different species may also influence the rate of lipid
autoxidation/degradation during the cooking and subsequent storage
periods (13). Thus, development of off-flavors and unpleasant odors
referred to as "warmed-over flavor" (7) depends primarily on the
degree of unsaturation of lipid components of meats and somewhat on
the level of iron-porphyrin materials present in the muscle.

Cured meats, on the other hand, have a distinct and pleasant
aroma which does not change significantly even after prolonged
storage at refrigerated temperatures. Antioxidant activity of
nitrite, the most important ingredient of the cure, may account for
this observation (14,15).

This study was undertaken to unravel some aspects of the flavor
of cooked meats and primarily to describe the effect of lipid
autoxidation on the flavor of cooked meats.

Meat Flavour Volatiles

Nearly 1000 compounds have so far been identified in the volatile
constituents of meat from beef, chicken, mutton and pork (6). The
largest number of volatiles has been determined in beef and these
were representative of most classes of organic compounds.
Hydrocarbons, alcohols, aldehydes, ketones, carboxylic acids,
esters, lactones, ethers, sulfur and halogenated compounds as well
as different classes of heterocyclic substances (Figure 1) namely
furans, pyridines, pyrazines, pyrroles, oxazol(in)es, thiazol(in)es,
thiophenes were present in cooked meat flavor volatiles as shown in
Table I. Many of these compounds are unimportant to the flavor of
meat and some may have been artifacts (16).

In our opinion, the predominant contribution to flavor seems to
come from sulfurous and carbonyl-containing volatiles. While many
of the sulfur-containing volatiles are known to have meaty aromas,
volatile carbonyl compounds generally are formed by lipid
autoxidation/degradation and do not possess meaty flavor notes.
However, it has been indicated that the carbonyl compounds are
responsible for the "chickeny" aroma of cooked chicken (17). Thus,
lipid autoxidation appears to yield the character impact compounds
for chicken (18).

TABLE I. Chemical Classes and Numbers of Volatile
Constituents of Meats

Class	Beef	Chicken	Mutton	Pork	Cured Pork
Hydrocarbons	123	71	26	45	4
Aldehydes	66	73	41	35	29
Ketones	59	31	23	38	12
Alcohols and Phenols	64	32	14	33	10
Carboxylic acids	20	9	46	5	20
Esters	33	7	5	20	9
Ethers	11	4	-	6	-
Lactones	33	2	14	2	-
Furans	40	13	6	29	5
Pyridines	10	10	16	5	-
Pyrazines	48	21	15	36	-
Pyrroles	4	6	1	9	1
Oxazol(in)es	10	4	-	4	-
Thiazol(in)es	17	18	5	5	-
Thiophenes	37	8	2	11	3
Other nitrogen compounds	6	5	2	6	2
Other sulfur compounds	90	25	10	20	30
Halogenated compounds	6	6	-	4	1
Miscellaneous compounds	5	2	-	1	11
Total	682	347	226	314	137

Although qualitatively many of the flavor volatiles present in meats from different species are similar, there are quantitative differences. In a recent review MacLeod (4) reported that mutton aromas have a high concentration of 3,5-dimethyl-1,2,4-trithiolane and thialdine (2,4,6-trimethylperhydro-1,3,5-dithiazine) and other sulfur compounds due to the presence of a higher concentration of sulfur-containing amino acids in mutton than in beef or pork. Furthermore, it was indicated that mutton aromas contain many alkyl-substituted heterocyclic compounds which may have been formed from the reaction of 2,4-alkadienals with NH_3 produced from thermal degradation of amino acids (19). These compounds are

$$NH_3 \ + \ CH_3-(CH_2)_4 \overset{\text{1) Heat}}{\underset{\text{2) }O_2}{\longrightarrow}} CH_3-(CH_2)_4$$

generally thought to be responsible for the roasted flavor notes and are associated with roasted rather than boiled meats (20). Again higher concentration of amino acids and lower content of sugars in mutton, as compared to beef or pork, accounted for this observation (21). These together with the presence of a larger number of carboxylic acids, and particularly branched chain saturated acids, and high levels of sulfur compounds may account for the rejection of mutton by certain consumers. In beef, mercaptothiophenes and mercaptofurans were significant contributors to its flavor and generally a lower contribution from the lipids on their overall flavor was observed.

Volatiles with Meaty Aromas

A total of 64 sulfur-containing compounds with meaty flavor characteristics have so far been identified in meat volatiles, from which only 7 were acylcic sulfides and thiols (Table II). Most of the sulfurous volatiles of cooked meats are organoleptically active. While at low concentrations present in cooked meats they exhibit a pleasant meaty aroma, at high concentrations their odor is objectionable. These compounds are generally produced from cysteine/cystine, glutathione, and thiamine upon the cooking of meat (Figure 2). Many of the sulfurous volatiles of meat with an active flavor note are heterocyclic in nature (22) and contain one or more sulfur atoms in their ring structure or as a side chain (Table II). A number of sulfur-containing compounds with meaty aromas have also been synthesized (23,24). These were generally thiols of substituted furans and thiophenes. Interestingly, none of these has been found in meat volatiles.

Volatiles with a meaty flavor note generally present in meats from different species are perhaps qualitatively the same, however, their quantities vary from one species to another (25,26). To date, only 13 non-sulfurous volatiles with meaty aromas have been identified in meats and some may indeed be artifacts (Table II).

Pyrazine

Pyridine

Pyrrole

Thiophene

Thiazole

3-Thiazoline

Oxazole

3-Oxazoline

Furan

Trithiolane

1,3,5-Trithiane

Tetrahydrothiophene

γ-Pyrone

Dithiazine

Quinoxaline

Figure 1. Chemical structures of some heterocyclic flavor
          volatiles of cooked meats.

$HSCH_2-CH-CO_2H$ Strecker $HSCH_2-C-H + NH_3 + H_2S + H_3C-C-H$
     | ⟶   ‖   ‖
    NH_2 Degradation   O   O
 (Cysteine)

Thiamine ⟶ Heterocyclic and Acyclic Sulfurous Compounds

Figure 2. Formation of products from breakdown of cysteine and
          thiamine.

Impact of Lipid Autoxidation/Degradation on Meat Flavor

The development of oxidative flavors and off-flavors is an important factor in acceptance or rejection of cooked meats. One of the important reactions involved in the formation of volatile compounds in meat, and meat products in general, is the autoxidation of unsaturated fatty acids. Phospholipid components of meats are generally rich in polyunsaturated fatty acids and hence are generally prone to autoxidation (27).

Autoxidation is described as having an initiation, a propagation and a termination step. The susceptability to autoxidation depends on the ability of fatty acids to donate a hydrogen atom during the propagation step. Thus, the carbon atoms adjacent to double bonds tend to donate a hydrogen atom leading to the formation of resonance-stabilized radicals. The primary products of lipid autoxidation are hydroperoxides and these are odorless in nature. However, upon decomposition of hydroperoxides, secondary products such as hydrocarbons, alcohols, ketones, and aldehydes are produced (Figure 3) and these influence the flavor of meat from different species. Depending on the composition of the fatty acids in lipids, the proportion of these oxidation products vary significantly. Furthermore, such products can themselves undergo further oxidation and decomposition, thus producing a large number of new products which include short-chain hydrocarbons, aldehydes, dialdehydes, epoxides, ketones, acids, alkyltrioxanes, dioxolanes, furans, as well as dimers and polymers.

Although autoxidation of lipids in foods is generally considered as unwanted, certain products of lipid autoxidation at low concentrations are necessary to the characteristic odor and aroma properties of meats from different species (8,9,28). Therefore, the concentration and relative abundance of these chemicals in meat volatiles determine whether they play a desirable or an undesirable role in flavor characteristics of cooked meats. Thus, the origin of flavor and off-flavors developments, which are somewhat species-specific, are perhaps the same. So, in freshly cooked meats the specific flavor of meat which is species-specific develops and progression of autoxidation results in the formation of undesirable warmed-over flavor in cooked meats upon storage.

Aldehydes and ketones, major secondary products of autoxidation are known to impart burnt, sweet, fatty, painty, metallic and rancid flavor notes to meats (6). Many aldehydes also have low odor and flavor thresholds and can be perceived at low concentrations (29, 30).

Malonaldehyde, a major product of autoxidation of polyunsaturated fatty acids is a very reactive substance and reacts with amino acids, proteins and other chemical substances present in meats. Its concentration is generally determined by the 2-thiobarbituric acid (TBA) test. Malonaldehyde may be used as an indicator for evaluation of the oxidative state of cooked meats. It has been reported that warmed-over flavor in beef is generally perceived when TBA number of cooked meats exceed numerical values of 0.5 to 1.0 (31). Malonaldehyde has also been implicated as having mutagenic and perhaps carcinogenic effects (32). Its presence further affects the rheological properties and texture of cooked meat products. Despite these, malonaldehyde has very little or no

TABLE II. Meat Flavor Volatiles with "Meaty" Aroma

Class of Compound	Number[a]	Compound	Example Flavor note[a]
Sulfides and Thiols, acyclic	7	Mercaptan, methylthioethane	meaty (1-5 ppb), onion
(Hydro)Furans with sulfur-containing side chain	17	Furfurylthiol, 5-methyl	meaty (0.5-1 ppb), sulfurous (>1 ppb)
Thiophenes	11	Thiophene-2-methyl-3-thiol	roast meat
Di and Trithiolanes	5	1,2,4-Trithiolane	roast meat
Trithianes	3	1,3,5-Trithiane,2,4,6-trimethyl	meaty
Thiazol(in)es	14	Thiazole	meaty, nutty, pyridine-like
Thialdines	4	Thialdine	meaty, roast beef
Pyrazine-furan sulfides	3	Furfurylthio-2-(-3-methyl) pyrazine	cooked meat (<1 ppb), coffee
Furans	2	Furan, 2-methyl	meaty, pleasant, slightly sulfurous, sickly
Oxazol(in)es	2	Oxazol, 2,4,5-trimethyl	boiled beef, nutty, sweet, green
Ketones	4	Cyclopentanone, 3-methyl	roast beef
Hydrocarbons	1	n-Octane	meaty
Miscellaneous	5	Thiophenol, 2-ethyl	meaty, burnt

[a]Ref. 4 and 6.

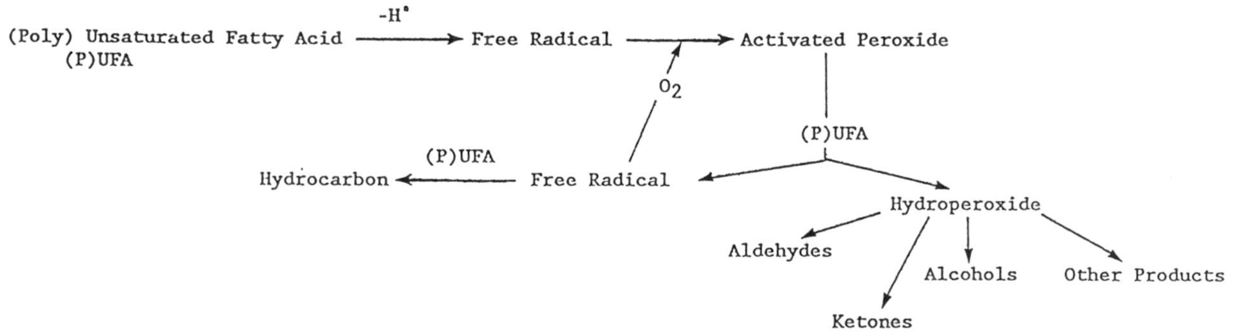

Figure 3. Mechanisms of lipid autoxidation and formation of products.

odor of its own and, in this respect, may have no effect on the
flavor of meat products.

Hexanal, on the other hand, is a predominant breakdown product
of lipid peroxidation of ω_6 fatty acids in meats.   Its influence on
the flavor of cooked meats, especially pork, mutton and chicken is
significant.   It has been described as having unpleasant, rancid,
green and pungent flavor notes (33,34).   Its content in cooked
ground pork was directly proportional to the amount of TBA-reactive
substances (TBARS) present (Figure 4).   Relative abundance of some
of the other aldehydes with respect to hexanal (arbitrarily set at
100) is given in Table III (unpublished results).   Similar results
were obtained for cooked ground chicken and mutton; however, beef
was somewhat less susceptible to autoxidation, and hexanal was less
abundant in the volatiles of beef.

Lipids or lipid breakdown products may also be involved in the
formation of 2-alkyl substituted heterocyclic compounds with roast
and/or fried flavor notes (19,35).   Therefore, lipid-derived
volatiles may have a special role in the development of flavor of
roasted and or barbecued meats.

Flavor Volatiles of Nitrite-Cured and Nitrite-Free Treated Meats

Nitrite is the unique ingredient of the cure due to its role in the
development of color, flavor, as well as oxidative and microbiol
stability to meats (36).   Each of these properties could be
duplicated, however, no single compound has been found with such
multifunctional properties.   Although nitrite is closely associated
with cured-meat aroma (14,15) the chemical changes that are
responsible for the unique flavor are not clearly understood (27).
A limited number of publications have appeared and a number of
volatile chemical constituents have been identified in cured pork.
Of particular interest is the work of Cross and Zeigler (37) in
which volatile constituents of both cured and uncured ham were
examined.   These authors reported that the concentration of
aldehydes, and especially pentanal and hexanal, was greatly reduced
in cured meats.   They also found that the volatiles from uncured
chicken and beef passed through a solution of 2,4-
dinitrophenylhydrazine had an aroma similar to that of cured ham.

Our own work has shown a great decrease in the concentration of
the volatiles in the cured, as compared to uncured, meats (Figure 5)
(38).   The concentration of aldehydes originally present in cooked
pork was reduced to $\leq$ 1% of their original quantities (Table III,
unpublished results).   However, we did not identify any new flavor
active compound which could have been responsible for the cured
flavor.   Lipid oxidation, as measured by TBA number, was almost
eliminated in cooked pork by adding nitrite at a level of 150 ppm
(39).   Furthermore in preliminary evaluations, our untrained
panelists were unable to differentiate amongst the flavor of
nitrite-cured meats prepared from beef, chicken, mutton and pork
(unpublished results).

It may then be reasonable to postulate that meat on cooking
acquires a characterisitic species flavor which is caused by the
volatile carbonyl compounds formed by oxidation of its lipid
components.   Due to the strong antioxidant activity of nitrite,
however, such oxidation products are either absent or are present

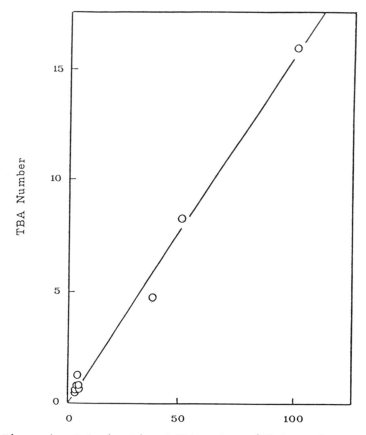

Figure 4.   Relationship of TBA numbers (35 days of storage) with
            relative hexanal content (2 days of storage).
            (Reproduced with permission from Ref. 42. Copyright
            1987. Can. Inst. Food Sci. Technol. J.).

TABLE III: Some Aldehydes and their Relative Abundance
             in Cooked Ground Pork

| Compound | Relative Abundance | | Odor Threshold |
	Uncured	Cured	ppb
Hexanal	100	7.0	4.5
Pentanal	31.3	0.5	-
Heptanal	3.8	<0.5	3.0
Octanal	3.6	<0.5	0.7
Nonanal	8.8	0.5	1.0
2-Octenal	2.0	-	3.0
2-Nonenal	1.0	-	0.08
2-Decenal	1.1	-	0.3
2-Undecenal	1.4	0.5	-
2,4-Dodecadienal	1.1	-	-

only in minute quantities in cooked cured products. Hence, it may be justified to assume that the flavor of nitrite-cured meats is actually the basic and natural flavor of meat from different species without being influenced by the overtones derived from autoxidation/degradation of their lipid components. However, there is a possibility that cured-meat flavor may have indeed been formed via a mechanism which is unrelated to lipid peroxidation.

In our search for nitrite alternatives, as far as flavor and oxidative stability is concerned, we examined the effect of commonly used adjuncts in meat curing, as well as a large number of antioxidant/sequesterant systems (40-42). In particular, the effect of sodium ascorbate (SA) and sodium tripolyphosphate (STPP) on the oxidative state of cooked meats was studied. These additives lowered the TBA numbers by a factor of about 2 and 4, respectively (Table IV). When used in combination, a strong synergism was observed. Furthermore, an increase in the concentration of SA and/or STPP resulted in a decrease in the TBA values as depicted in Figure 6 (43). Addition of 30 ppm of butylated hydroxyanisole (BHA) or tert-butylhydroquinone (TBHQ) further reduced the TBA numbers and in fact the latter values were even lower than those obtained for meats treated with sodium nitrite (Table IV) (41).

The content of hexanal (Table III) and other lipid oxidation-derived flavor compounds, of meats treated with STPP and/or SA was similarly reduced and the spectrum of flavor volatiles of meats was simplified (Figure 6C), and was similar to that from nitrite-cured meat (Figure 6B). The data for hexanal content are in agreement with Cross and Zeigler's findings (37). Both SA and STPP alone lowered the amount of hexanal in the meat and when used in combination, a strong synergistic effect was observed, thus lending support to our previous findings on TBA values (see above). Although additions of BHA, TBHQ or nitrite, at a level of 30 ppm, had a further effect in reducing the amount of hexanal, the major effect was due to the combination of SA and STPP (42). Furthermore, flavor acceptability of nitrite-free treated samples was not significantly different from that of nitrite-cured meat, as determined by our untrained panelists (Table IV). Therefore, it appears that nitrite may not be an essential ingredient for the development of characterisitc flavor of (certain) cured meat products.

Future Research Needs

As a result of the availability of sophisticated instrumentation and separation techniques some remarkable progress has already been made in meat flavor research and this trend is expected to continue. Although a variety of factors are known to affect the development of meat flavor, no single compound/group of compounds, or factor has yet been found that could play the principle role and the true chemical nature of meat flavor, and particularly species differentiation, is not fully understood. Most importantly very little is known about the origin of cured-meat flavor. The curing process seems to simplify the composition of the volatile constituents and eliminates the overtones related to species-specific flavor notes. Thus, work in this area would have a major impact in meat-flavor research and may prove to be extremely

TABLE IV.   TBA Numbers, Hexanal Content, and Sensory Scores of Meat [a]

No.	Meat System Additives (ppm)[b]	TBA Number	Hexanal	Sensory Score
1	No Additives	15.93(4.72)	100	2.8
2	(1) + SA(550)	8.30(1.35)	50.1	-
3	(1) + STPP(3000)	4.79(0.21)	38.0	5.1
4	(2) + STPP(3000)	1.33(0.20)	4.1	-
5	(4) + BHA(30)	0.22(0.20)	1.6	5.3
6	(4) + TBHQ(30)	0.27(0.21)	3.0	5.6
7	(4) + NaNO$_2$(30)	0.56(0.30)	2.1	-
8	(4) + NaNO$_2$(150)	0.43(0.25)	2.0	5.7
9	(2) + NaNO$_2$(150)	0.39(0.28)	2.8	-

[a]From Ref. 42 and unpublished data. Cooked meat were stored at 4°C for 2 days
for hexanal, sensory score and TBA numbers (in brakets) determination. Other
TBA numbers were determined after 35 days of storage.
[b]All meat samples contained 2% NaCl and 1.5% sucrose. The additives were sodium
ascorbate: SA, sodium tripolyphosphate: STPP, butylated hydroxyanisole: BHA,
tert-butylhydroquinone: TBHQ, and sodium nitrite, NaNO$_2$.

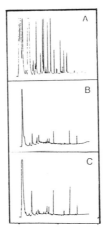

Figure 5. Gas chromatograms of volatiles of meats with A: no additive, B: sodium ascorbate (550 ppm) and sodium nitrite (150 ppm), and C: sodium tripolyphosphate (3000 ppm), sodium ascorbate (550 ppm) and butylated hydroxyanisole (30 ppm).

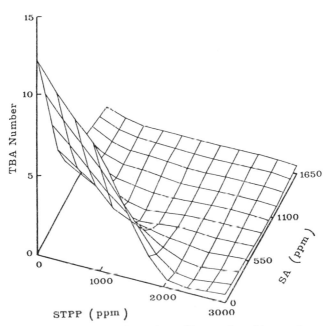

Figure 6. 3-D plot showing the effect of sodium tripolyphosphate (STPP) and sodium ascorbate (SA) on the TBA numbers of cooked pork after 4 weeks of storage at 4°C. (Reproduced with permission from Ref. 43. Copyright 1987. Can. Inst. Food Sci. Technol. J.).

fruitful.   Identification of possible flavor-active components present in extremely low concentrations may be rewarding.   Use of non-degradative methods of isolation such as super-critical extraction and identification of flavor compounds by HPCL and HPLC-MS methods are highly recommended.

Furthermore, growing needs for creation of novel fast foods with characteristic aromas, consumers awareness and their desire for natural flavors may dictate, in part, the future direction that meat flavor research may pursue.   Creation of simulated meat flavors with acceptable quality characteristics may satisfy a growing and discriminating population of health-conscious consumers.   The role of lipids and lipid-derived aroma volatiles and creation of species-specific meat flavors deserves further investigation.

Literature Cited

1.   Crocker, E.C. 1948.   Food Res. 1948, 13, 179-183.
2.   Bender, A.E.; Ballance, P.E.   J. Sci. Food Agric. 1961, 12, 683-687.
3.   Hornstein, I.; Teranishi, R.   Chem. Eng. News 1967, 45(15), 92-108.
4.   MacLeod, G. 1986.   In Developments in Food Flavours; Birch, G.G.; Lindley, M.G., Eds.; Elsevier Applied Science:London and New York, 1986; pp 191-223.
5.   Maillard, L.C. Compt. Rend. 1912, 154, 66-68.
6.   Shahidi, F.; Rubin, L.J.; D'Souza, L.A.   CRC Crit. Rev. Food Sci. Nutr. 1986, 24, 141-243.
7.   Pearson, A.M.; Love, J.D.; Shorland, F.B.   In Advances in Food Research; Chichester, C.O.; Mrak, E.M.; Stewart, G.F., Eds; Academic Press: New York, 1997; Vol. 23, pp 1-74.
8.   Mottram, D.S.; Edwards, R.A.; MacFie, J.   J. Sci. Food Agric. 1982, 33, 934-944.
9.   Mottram, D.S.; Edwards, R.A.   J. Sci. Food Agric. 1983, 34, 517-522.
10. Wong, E., Nixon; L.N.; Johnson, C.B.   J. Agric. Food Chem. 1975, 23, 495-498.
11. Wong, E., Johnson, C.B.; Nixon, L.N.   N.Z.J. Agric. Res. 1975, 18, 261-266.
12. Gower, D.B.; Hancok, M.R.; Bannister, L.H.   In Biochemistry of Taste and Olfaction; Cagan, R.H.; Kave, M.R., Eds., Academic Press:New York, 1981; pp. 7-31.
13. Younathan, M.T.; Watts, B.M.   Food Res. 1959, 24, 728-734.
14. MacDonald, B.; Gray, J.I.; Kakuda, Y.; Lee, M.L.   J. Food Sci. 1980, 45, 889-892.
15. MacDonald, B.; Gray, J.I.; Kakuda, Y.; Lee, M.L.   J. Food Sci. 1980, 45, 889-892.
16. MacLeod, G.; Seyyedian-Ardebili, M.   CRC Crit. Rev. Food Sci. Nutr. 1981, 14, 309-437.
17. Minor, L.J.; Pearson, A.M.; Dawson, L.E.; Schewigert, B.S.   J. Food Sci. 1965, 30, 686-696.
18. Harkes, P.D.; Begemann, W.J.   J. Amer. Oil Chem. Soc. 1974, 51, 356-359.
19. Buttery, R.G.; Ling, L.C.; Teranishi, R.; Mon, T.R.   J. Agric. Food Chem. 1977, 25, 1227-1229.

20. Watanabe, K.; Sato, Y. Agric. Biol. Chem. 1971, 35, 756-763.
21. Baines, D.A.; Mlotkiewicz, J.A. In Recent Advances in the Chemistry of Meat; Baily, H.J., Ed.; Royal Soc. Chem. London, 1984; pp. 119-64.
22. Nurstein, H.E. In Progress in Food and Nutrition Science; Eriksson, C., Ed.; Pergamon Press:Oxford, 1981; Vol. 5, pp. 491-496.
23. van den Ouwenland, G.A.M.; Peer, H.G. J. Agric. Food Chem. 1975, 23, 501-505.
24. Chang, S.S.; Peterson, R.J. J. Food Sci. 1977, 42, 298-305.
25. Hornstein, I.; Crowe, P.F. J. Agric. Food Chem. 1960, 8, 494-498.
26. Hornstein, I.; Crowe, P.F. J. Agric. Food Chem. 1963, 11, 147-149.
27. Gray, J.I.; Pearson, A.M. In Advances in Food Research; Chichester, C.O.; Mrak, E.M.; Scheweigart, B.S., Ed; Academic Press: New York, 1984; Vol. 29, pp 1-86.
28. Herz, K.O.; Chang, S.S. In Advances in Food Research; Chichester, C.O.; Mrak, E.M.; Stewart, G.F., Eds.; Academic Press:New York, 1970; Vol. 18, pp. 1-83.
29. Guadagni, D.G.; Buttery, R.G.; Okano, S. J. Sci. Food Agric. 1963, 14, 761-765.
30. Frazzalari, F.A. Compilation of Odor and Taste Threshold Value Data; American Society for Tasting and Materials: Philadelphia, 1978.
31. Tarladgis, B.G.; Walls, B.M.; Younathan, M.T., Dugan, Jr., L.R. J. Amer. Oil Chem. Soc. 1960, 37, 44-48
32. Shamberger, R.J.; Andreone, T.L.; Willis, C.E. J. Natl. Cancer Inst. 1974, 53, 1771-1773.
33. MacLeod, G.; Coppock, B. J. Agric. Food Chem. 1976, 24, 835-843.
34. Persson, T.; von Sydow, E. J. Food Sci. 1973, 38, 377-385.
35. Tang, J.; Jin, Q.Z.; Shen, G.-H.; Ho, C.-T.; Chang, S.S. J. Agric. Food Chem. 1983, 31, 1287-1292.
36. Shahidi, F.; Rubin, L.J.; Diosady, L.L.; Chew, V.; Wood, D.F. Can. Inst. Food Sci. Technol. 1984, 17, 33-37.
37. Cross, C.K.; Ziegler, P. J. Food Sci. 1965, 30, 610-614.
38. Yun, J. M.A.Sc. Thesis, University of Toronto, Toronto, 1984.
39. Shahidi, F.; Rubin, L.J.; Wood, D.F. J. Food Sci. 1987, 52, 564-567.
40. Shahidi, F.; Rubin, L.J.; Diosady, L.L.; Kassam, N.; Li Sui Fong, J.C. Food Chem. 1986, 21, 145-152.
41. Shahidi, F.; Rubin, L.J.; Wood, D.F. J. Food Sci. 1987, 23, 151-157.
42. Shahidi, F.; Yun, J.; Rubin, L.J.; and Wood, D.F. Can. Inst. Food Sci. Technol. J. 1987, 20, 104-106.
43. Yun, J.; Shahidi, F.; Rubin, L.J.; Diosady, L.L. Can. Inst. Food Sci. Technol. J. 1987, 20, 246-251.

RECEIVED September 29, 1988

Chapter 16

New Trends in Black Truffle Aroma Analysis

T. Talou, M. Delmas, and A. Gaset

Laboratoire de Chimie des Agroressources, Ecole National Supérieure de Chimie, I.N.P.T., 118, route de Narbonne 31077 Toulouse Cedex, France

The volatile compounds in the atmosphere of cold stored Black Perigord Truffles (Tuber Melanosporum) were adsorbed onto a Tenax trap by means of a vacuum pump. The efficiency of the sampling method was sensorially validated. The volatiles eluted from the trap by heat desorption were analysed by capillary gas chromatography - mass spectrometry. A total of 26 compounds was identified . Their contribution to the final aroma impression was discussed.

The isolation of volatile flavor compounds from foods represents a major problem in analytical studies of food flavor. Headspace methods are especially attractive since they are rapid, simple and measure what is typically presented to the nose.

We have previously used a Dynamic Headspace method, optimized with Experimental Design (1), for the isolation of the volatiles from Black Perigord Truffles (Tuber Melanosporum). Truffles are underground mushrooms that grow in symbiosis with certain trees, especially oaks. One finds them in several regions of Europe, particularly in France, where their flavor is very much appreciated by gourmets. Our studies, carried out on truffle flesh (2,3), and entire truffles (Talou, T. et all, J. Sci Food Agric., in press), allowed us to identify the major volatile compounds. Due to the low sampling weight, however, the method appeared to be ineffective for isolating compounds extremely low in concentration.

The atmosphere of cold stored Black Truffles is particularly rich in volatile compounds which impart the truffle aroma. We therefore developed a modified gas headspace sampling procedure for their isolation.

The aims of this study were to evaluate the efficiency of our sampling method for trapping volatile components important to the aroma of Black Perigord Truffles, and the analysis of minor volatiles via their identifcation with capillary gas chromatography-mass spectrometry.

0097–6156/89/0388–0202$06.00/0

Experimental

Plant material Fresh Black Perigord Truffles (Tuber Melanosporum) for analysis were purchased by Pebeyre Ltd., a company specializing in truffle marketing. Collected essentially in the South East of France, they were fully ripe (4) and released their characteristic aroma. Received the day after gathering in wicker-baskets, truffles were hand-picked and then stocked in a cold storage of 15 m^3.

Volatile collection was generally begun the day after receiving the packages. 1000 kg of unbrushed truffles were in storage when atmosphere capture was carried out. Three isolations were carried out in duplicate during February 1987.

Preparation of Tenax tubes and Blank Tenax chromatogram Tenax GC (60-80 mesh) was conditionned by heating 10 g at 250°C for 24 h in a dry Helium flow of 50 mL/min. Conditioned Tenax (0,2 g) was packed into a 6 cm long steel tubes (6 mm o.d., 4 mm i.d.) and plugged with glass wool. A blank Tenax gas chromatogram was obtained by heat desorbing the tube into a gas chromatograph (GC) column using the procedure described under Heat Desorption of Aroma volatiles section. Additional cleaning of Tenax tubes was standardized by heat desorption at 200°C for 30 min in the oven of the D.C.I. System.

Headspace sampling method An idealized diagram of the sampling system is shown in Figure 1. It consisted of a vacuum pump, a needle valve, a flow meter and 2 auxiliary Tenax tubes. Connections were made by Swagelock Teflon unions. Cold storage atmosphere was passed through the two "in-serie" ambient Tenax tubes by applying suction to the outlet of the tube by means of the vacuum pump.

Preliminary experiments, the results of which are summarized briefly under Results and Discussion, led to the following optimized sampling method. Volatile collection was carried out for 30 min at the cold storage temperature (4°C) and with a vacuum pump flow rate of 50 mL/min. Then, the sampled Tenax tubes were removed, capped and ready for GC analysis the same day.

Heat Desorption of Aroma volatiles The gas chromatographic method employed a D.C.I. System (Desorption - Concentration - GC Introduction), available from Delsi Instruments (5). A schematic diagram of this apparatus is given in Figure 2.

One of the auxiliary Tenax tubes (1) was placed inside the desorption oven (2), located upstream of the fixed Tenax trap (3) (0,2 g Tenax GC, 60-80 mesh, packed into a 7 cm by 2 mm i.d. stainless steel tube). At controlled temperature (100°C) and low pressure (1,5 psi) the oven was flushed by a 25 mL/min flow of Helium for 10 min. Desorbed and diluted in scavenging gas (a), the volatiles were then concentrated and trapped in the Tenax trap, cooled to -30°C by circulation of liquid nitrogen. By switching a rotary valve (4), carrier gas (b) flowed through the trap (backflush) and towards the GC column (6). By rapid thermal desorption at 240°C, aroma volatiles were directly transferred onto the GC column. Figure 3 shows the chronological sequence for cryogenic volatile adsorption and thermal desorption, and chromatographic separation.

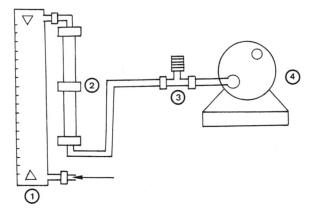

Figure 1.   Schematic diagram of the sampling apparatus
1) flowmeter; 2) auxiliary Tenax tube; 3) needle valve; 4) vacuum
pump.

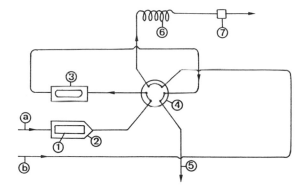

DESORPTION - CONCENTRATION

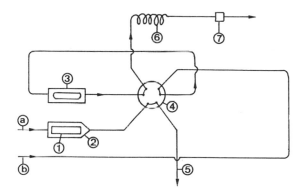

INTRODUCTION

Figure 2.   Schematic diagram of the D.C.I. system
a) scavenger gas; b) carrier gas
1) auxiliary Tenax tube; 2) desorption oven; 3) fixed Tenax trap;
4) switching valve; 5) gas vent; 6) GC column; 7) mass spectrometer

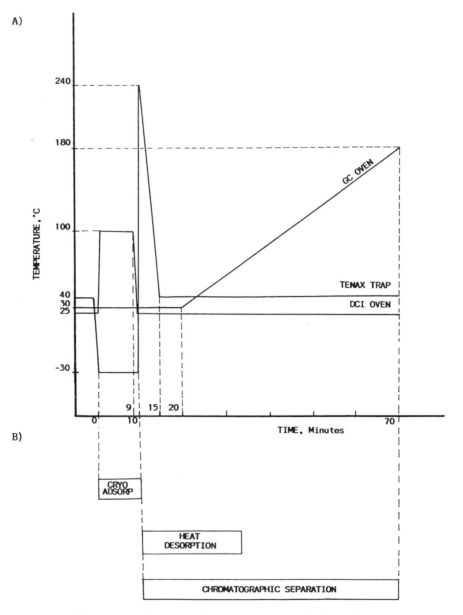

Figure 3. Chronological sequences of GC analysis
A) temperature sequence; B) time sequence.

Combined Capillary Gas Chromatography - Mass Spectrometry A GIRDEL
31 gas chromatograph equipped with a D.C.I. System was coupled by
means of a glass lined tubing interface to a NERMAG R10-10B
quadrupole mass filter spectrometer. The system was connected on-
line to a dedicated data processing system consisting of a Digital
Equipment Co. PDP8 Computer, using SIDAR software including the
library of mass spectral data NIH/EPA (6).
    The capillary column used was a 50 m X 0,32 mm (i.d.) fused
silica UCON 75H 90000 wall coated column. The column oven
temperature was programmed from 30 to 180°C at a rate of 3°C/min
with a 10-min post injection hold and a 10-min hold at a final
limit. Column inlet pressure of Helium was 7 psi and the split 30
mL/min.
    Significant Mass Spectrometry operational parameters were as
follows : ionization voltage, 70 eV; ionization current, 200 uA;
source temperature, 200°C; electron multiplier tension, 1,9 kV;
integration time, 1 ms/u.m.a.. For optimum sample transfer, an
interface temperature of 200°C was adopted.

Reagents Helium was purified by passage through charcoal,
molecular sieve and an Oxisorb trap. Tenax adsorbant was obtained
from Alltech Assoc. Inc. Chemical compounds for comparisons of mass
spectra and GC retention times were obtained from commercial sources
(Aldrich). Sec-butyl and isopropyl formates were synthesized
according to Dimicky (7).

Sensory Validation of Sampling and GC Techniques The sensory
evaluation was carried out by a panel of three judges (employees of
Pebeyre Ltd.). For this study, an external odor port was attached to
the gas vent (5) of the D.C.I. System and the rotary valve (4) was
not switched (analysis using the Desorption - Concentration mode).
Thus, after the thermal desorption of the volatiles from the
trap, the rotary valve was positioned so that the unresolved aroma
isolate went to our sniffing port. The response was measured as
similarity or dissimilarity to Black Truffle aroma.

Sensory Evaluation of the contribution of identified compounds to
the aroma In order to identify those compounds mainly contributing
to the characteristic flavor of the Black Truffle, the odor of the
individual components of the headspace analysis were tested by a
panel of eight judges (trained in sensory evaluation of truffles).
The compounds tested were diluted in vegetable oil, in a range of
concentrations from 30 to 300 ppm. 5 point scales (5 = exceptionally
good full truffle aroma, 1 = not different from solvent) were used
for flavor imitation and intensity.

Results and Discussions

Optimization of the sampling and analytical techniques Preliminary
trials were used to determine the optimal adsorbent. Tenax GC was
chosen due to its high affinity for organic compounds which it
adsorbs reversibly, and its relatively hydrophobic character. This
point is important in view of the large volume of water vapor
present in the cold storage atmosphere. For this same reason,

subambient traps appeared to be innefective due to the risk of physical blockage of the Tenax by solidification of atmospheric water vapor.

The optimization of the sampling conditions was necessary to avoid breakthrough from the Tenax, i.e. elution of volatiles partially through the trap during atmosphere capture. Indeed, sampling efficiency decreased with sampling time, breakthrough then occurring, but this was not significantly decreased by using lower sampling flow rates. Under our optimized sampling conditions, the heat desorption of the second Tenax trap which was nearest to the vacuum pump in the sampling system gave a blank chromatogram, showing that there were no losses of volatiles from the first trap during the capture.

The conditions of chromatographic analysis (desorption of volatiles from the Tenax tube and their adsorption on the Fixed Tenax trap) have also been optimized in order i) to obtain a total desorption of volatiles trapped on the auxiliary tube (confirmed by a blank chromatogram for a second heat desorption), ii) to avoid losses of volatiles during the adsorption phase (sensorially verified at the odor port).

One particular volatile isolate desorbed and assessed sensorially at our odor port was described as typical of Black Truffle, showing that the Tenax had adsorbed and desorbed (under the analytical conditions used) volatile components that impart Black Truffle aroma.

The aroma isolation method and heat desorption techniques were therefore validated.

Identification of volatiles compounds A typical total ion current chromatogram of the Tenax trapped Black Truffle (Tuber Melanosporum) volatiles is shown in Figure 4.

The compounds identified by GC-MS are listed in Table I, in order of elution from the GC column with their observed characteristic mass spectral data. The identification of these compounds is based on comparison of their respective mass spectra obtained with those stored in the NIH/EPA library and then with those of authentic compounds. Moreover, an additional search of published standard mass spectra to confirm the identity of unknown was undertaken (8).

The isolation of the major volatile compounds, i.e. low boiling point alcohols, and their relative aldehydes, 2 ketones and one major sulfur compound was consistent with previous reports (2,3). Although alcohols remained the major constituents of headspace volatiles, dimethyl sulfide was present in greater proportion than in our earlier studies.

The similarity of the chromatographic profiles of headspace volatiles of brushed (1,2) and unbrushed truffles (obtained in this study) allowed us to give scientific support to the informal subjective observation that unbrushed and brushed truffle aroma are not significantly different.

Nevertheless, we noticed that the relative concentration of alcohols increased during the period of cold storage. But the experts (employees of Pebeyre Ltd.) reported that the aroma of the analysed truffles were not really altered by this modification of

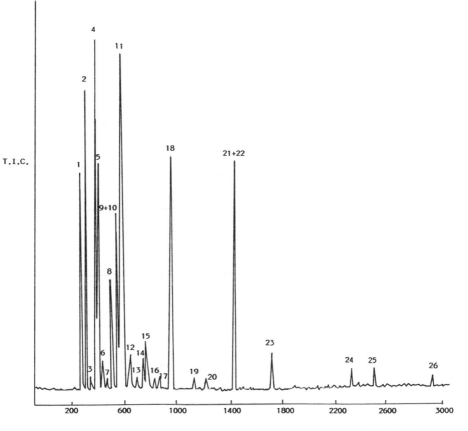

Figure 4. Reconstructed capillary GC-total ion current chromatogram of the Tenax trapped Black Truffle (Tuber Melanosporum) headspace volatiles.

Table I. Volatiles identified in atmosphere of cold storage
for Black Truffles (Tuber Melanosporum)

peak number (a)	compound	mass spectral data (b)	identification GC,MS
1	acetaldehyde	29,44,43,42	GC,MS
2	dimethyl sulfide	42,62,45,46	GC,MS
3	propanal	29,28,27,58	GC,MS
4	acetone	43,58,42,39	GC,MS
5	2-methylpropanal	43,41,72,27	GC,MS
6	isopropyl formate	45,42,43,73	GC,MS
7	ethyl acetate	43,29,61,45	MS
8	2-butanone	43,72,29,27	GC,MS
9	2-methylbutanal	41,57,29,58	GC,MS
10	3-methylbutanal	44,41,43,29	GC,MS
11	ethanol	31,45,46,29	GC,MS
12	sec-butyl formate	45,41,56,29	GC,MS
13	chloroform	83,85,47,48	MS
14	2-butanol	45,31,59,29	GC,MS
15	1-propanol	31,29,42,27	GC,MS
16	toluene	91,92,65,51	MS
17	dimethyl disulfide	94,45,79,46	MS
18	2-methyl-1-propanol	43,33,42,41	GC,MS
19	ethyl benzene	91,106,51,28	MS
20	xylene	91,106,105,77	MS
21	2-methyl-1-butanol	41,29,57,56	GC,MS
22	3-methyl-1-butanol	41,29,42,55	GC,MS
23	anisole	108,78,65,39	GC,MS
24	acetic acid	45,43,60,28	MS
25	benzaldehyde	105,106,77,51	GC,MS
26	2-formyl thiophene	111,112,39,45	MS

a) The peak numbers correspond to numbers in Figure 4; b) The four
most intense peaks are reported.

concentrations. This result showed that a short period of cold storage did not affect the organoleptic quality of truffles.

The major advantage of the sampling technique developed, was that some trace chemicals could be trapped and described for the first time as Black Truffle aroma constituents. In particular, some compounds, important flavor contributors, generally appearing in small concentrations, such as benzaldehyde, propanal, ethyl acetate, anisole or dimethyl disulfide - previously identified in Shiitake mushrooms (9) - could be characterized. This was also the case for three aromatic compounds, toluene, xylene and ethyl benzene, well known as raw vegetable constituents (10). In addition, two aliphatic esters, isopropyl and sec-butyl formates, and one cyclic sulfur compound (2-formyl thiophene) previously reported respectively in plums and apples (10) and in coffee and bread products (11) were identified.

Chloroform was certainly an artifact due to treatment of the soil where the truffles were harvested.

Relative odor contribution of components   Considered individually, none of the compounds identified duplicated the flavor of the Black Truffle. However, according to the judges, dimethyl sulfide was determined to have a great importance to the final aroma impression. In the same way, 2-butanone and anisole were found to be responsible for an off-flavor characteristic of over-mature truffles.

When mixtures of two and/or three compounds were tested, the panel scores remained low except for the mixture of dimethyl sulfide/2-methylbutanal.

A synosmic effect, which has been defined by Schutte (12) as being due to two or more compounds combining together to give a flavor sensation very different from that of any single component, seemed evident. This was confirmed by the high score obtained by a mixture of all compounds. This oil solution was judged to be very similar to fresh black truffle aroma by the panel (Talou, T. Ph.D. Thesis, in preparation): this indicated that the typical flavor for black Perigord Truffle was the result of an integrated response to the contribution of all the identified compounds.

An industrial application of this result was the elaboration of the first Nature-Identical Black Truffle aromatizer (13). This product was tested with success during one year by well known french cooks in their restaurant. Culinary tests confirm the good organoleptic quality of the flavored products obtained, both for cold or hot dishes.

Conclusions

The headspace sampling technique developed in the present study to collect volatiles from cold stored Black Truffles performed adequately. Indeed, the aroma isolate obtained was described as typical, and 11 minor compounds could be described for the first time as Black Truffle aroma constituents. Moreover, these results allowed the formulation of the first Nature-Identical Black Truffle aromatizer.

Acknowledgments This research was conducted in partial fulfillment of the requirements of the Ph.D. degree by T. Talou. The authors thank Dr. O. Morin of the department of Analytical Chemistry of ITERG (Bordeaux, France) for the GC-MS analysis, and Ms. Suzanne Harington and Sylvie Bourdarie for their contributions in the preparation of this paper. This study was a part of a Research Program sponsored by Pebeyre Ltd. (Cahors, France), the French Ministry of Agriculture and the Midi-Pyrénées Regional Council.

Literature Cited

1. Talou, T.; Rigal, L. and Gaset, A. Proc. 1st Colloquium Chemio-metricum Mediteraneum, 1987, p18.
2. Talou, T.; Delmas, M. and Gaset, A. J. Agric. Food Chem., 1987, 35, 774.
3. Talou, T.; Delmas, M. and Gaset, A. Proc. 5th International Flavor Conference, 1987, p 14.
4. Kulifaj, M. Ph.D. Thesis, University of Paul Sabatier, Toulouse, France, 1984.
5. Gregoire, J. and Samoun, A.M. Proc. 33rd Pittsburg Conference and Exposition on Analytical Chemistry and Applied Spectroscopy, 1982, p 768.
6. Heller, S.R. and Milne, G.W.A., EPA/NIH Mass spectra data base ; US Government printing office: Washington, DC, 1978; Vol 1-4.
7. Dimicky, M. Organic Preparations and Procedures Int. 1982, 14, 177
8. Stenhagen,E.; Abrahamsson, S. and MacLafferty, K.W. Registry of Mass Spectra data ; Willey: New York, 1974.
9. Chu-Chin, C.; Su-Er, L.; Chung-May, W. and Chi-Tang,H. ACS Symp. Ser.317 ; Parliment, T.H. and Croteau, R. Ed:,ACS Washington DC, 1986 ; p 14.
10. Teranishi, R. Flath, R.A. and Sugisawa,H. Flavor Research Recent Advances ; Marcel Dekker : New York, 1981.
11. Fenaroli, G. HandBook of Flavor Ingredients, 2nd Ed; CRC : Cleveland,OH, 1975 ; Vol 2.
12. Schutte, L.,ACS Symp. Ser.26 ; Charalambous, G. and Katz, I., Ed:, ACS : Washington DC, 1976 ; p 96.
13. Talou, T.; Delmas, M.; Gaset, A.; Montant, C. and Pebeyre, P.J. European Patent 87201148.1, 1987 ; Japaneese Patent 62.183242, 1987 ; U.S. Patent 077.276, 1987.

RECEIVED August 8, 1988

Chapter 17

Fresh Tomato Volatiles

Composition and Sensory Studies

Ron G. Buttery, Roy Teranishi, Robert A. Flath, and Louisa C. Ling

Western Regional Research Center, Agricultural Research Service, U.S. Department of Agriculture, Albany, CA 94710

Analysis of the vacuum volatile constituents of fresh tomatoes was carried out using capillary GLC-MS and packed column GLC separation with infrared, NMR and CI-MS analysis. Evidence was obtained for the presence of the unusual components β-damascenone, 1-nitro-2-phenylethane, 1-nitro-3-methylbutane, β-cyclocitral and epoxy-β-ionone. A method for the quantitative analysis of the volatile aroma components in fresh tomato has been improved and applied to fresh tomato samples. The quantitative data obtained have been combined with odor threshold data to calculate odor unit values (ratio of concentration / threshold) for 30 major tomato components. These calculations indicate that the major contributors to fresh tomato aroma include (Z)-3-hexenal, β-ionone, hexanal, β-damascenone, 1-penten-3-one, 3-methylbutanal, (E)-2-hexenal, 2-isobutylthiazole, 1-nitrophenylethane and (E)-2-heptenal.

The authors are carrying out a continuing study to try to obtain a better chemical definition of fresh tomato flavor and aroma. Studies to develop and apply quantitative methods to the analysis of fresh tomato volatiles have been recently carried out by some of the authors (1,2). Besides the known major compounds a number of compounds were detected in the gas liquid chromatography (GLC) analysis which had spectral data unlike that of any of the 400 compounds previously reported as tomato volatiles (cf. 3). As these compounds occurred in reasonable amounts in fresh tomato it seemed necessary to determine their identities in order to give a satisfactory quantitative picture of fresh tomato volatiles. It also seemed desirable to determine the odor threshold of these compounds to have a better understanding of their probable contribution to tomato aroma.

EXPERIMENTAL

Materials. Tomatoes were grown on experimental and commercial fields near Davis, California during the summer of 1987. Tomato

breeding lines used included E6203, FM785, GS-12 (Goldsmith-12), Lassen, XPH5498 and others.   Freshly picked vine ripe tomato samples were stored at room temperature under normal lighting and used within 3-5 days.
Freshly distilled diethyl ether and saturated $CaCl_2$ solution were prepared as previously described (2).

Isolation of Volatiles from Tomato Condensate. Condensate from commerical tomato paste production using vacuum concentration was stored at $5^{\circ}C$ in the dark and used within a few days.
Extraction was carried out with a laboratory built 40 liter liquid-liquid pyrex extraction apparatus.   The tomato condensate was first extracted with pentane for 24 hours and then with diethyl ether for 24 hours.   The solvent was removed by distillation from a warm water bath using a Vigreux distillation column.   The concentrates from a number of batches of tomato condensate were combined.   Ethyl antioxidant 330 (1,3,5-trimethyl-2,4,6-tris-[3,5-ditertbutyl-4-hydroxybenzyl]-benzene; ca. 10 ppm) was added and the concentrate stored at $-20^{\circ}C$.

Packed Column GLC Separation of Components.   The concentrate from above was first separated into two main fractions by micro-distillation.   These were Fraction A, b.p. 25-38$^{\circ}C$ at 0.1 mm Hg   (88% of original concentrate) and Fraction B, b.p.$>$ 38$^{\circ}C$ at 0.1 mm Hg (residue, 12% of concentrate).
Components were isolated from these distillation fractions using consecutive GLC separation first on a 10 m x 1.3 cm o.d. aluminum column packed with 60-80 mesh Chromosorb G coated with 5% Silicone SF96(50) followed by further resolution on a 3 m x 0.95 cm o.d. Pyrex glass column packed with 60-80 mesh Chromosorb G coated with 1% Carbowax 20-M.   A specially designed glass collector packed with Pyrex glass wool was used for the 10 m column (the trap was centrifuged to separate the component from the glass wool) and 10 cm long by 3 mm o.d. pyrex tube collectors for the 3 m column. During collection the traps were cooled with dry ice.   The collected samples were sealed in the 3 mm tubes and stored at $-20^{\circ}C$.

Infrared and NMR Spectra. Infrared spectra were measured as thin films using ultramicro salt plates with a Perkin Elmer Model 197 instrument.   Proton nuclear magnetic resonance ($^{1}HNMR$) spectra were measured as solutions in $CDCl_3$ using a Nicolet NTC 200FT spectrometer.

Gas-Chromatography Mass Spectral (GC-MS) Analyses.   Several different studies were carried out.   The main study was done using a Finnigan MAT 4500 series quadrupole mass spectrometer and a 60 m x 0.32 mm i.d. DB-1 bonded fused silica capillary GLC column. The column was programmed from 25-250$^{\circ}C$ at 4° per minute with an inlet pressure of 14 psi.   Chemical ionization (CI) mass spectra on some of the components were also obtained using a VG Micromass 70/70 mass spectrometer with isobutane as the reactant gas.

Isolation of Volatiles for Quantitative Studies.   The method used
was essentually the same as that described previously (2).   The
whole tomato sample (100g at 25°C) of pieces cut from 3-4
different tomatoes was blended for 30 seconds (using a Waring
blender with blades rotating at 13670 rev/min) .   The mixture was
allowed to stand at room temperature for 180 seconds longer and
then saturated CaCl$_2$ solution (100ml) added and the mixture
blended for 10 seconds.   A standard solution (5.0 ml)
containing 20.0 ppm 2-octanone, 20.0 ppm 3-pentanone and 5.0 ppm
anethole in water (the standard solution was stored at 5°C in the
dark) was then added and the mixture blended for 10
seconds.   The resultant mixture was then poured into a 1 L flask
containing an efficient magnetic stirrer.   Purified air (3L/
minute) was then led into the flask and passed over the vigorously
stirred mixture (at 25°C) and out of the flask through a Tenax
trap (14 cm long by 2.2 cm i.d.; 10 g).   All connections were
either Pyrex glass or Teflon.   The isolation was carried out for
60 minutes and the trap removed and eluted with 100 ml of diethyl
ether.   The ether extract was concentrated to ca. 50 µl using a
warm water bath and Vigreux distillation column.   The Tenax trap
was reactivated by passing a stream of purified nitrogen through it
at 200°C for 1 hour.

Authentic Samples.   Authentic samples of identified compounds were
obtained from reliable commercial sources or synthesized by
established methods.   All samples were purified by GLC separation
and their identities verified by mass or infrared spectrometry.
1-Nitro-3-methylbutane and 1-nitro-2-phenylethane were synthesized
according to the method of Kornblum et al.(4) by the reactions of
1-bromo-3-methylbutane and 2-phenylethylbromide with sodium nitrite
in dimethylformamide and urea.

Odor Threshold Determinations.   These were carried out on samples
purified by gas chromatographic separation using methods previously
described (1) with a panel of 16 to 20 judges.

RESULTS AND DISCUSSION

Three main approaches were applied to fresh tomatoes.   The first
approach was a qualitative one.   It was aimed at the further
identification of important aroma compounds.   The second approach
was designed to develop better methods for the quantitative
analysis of important tomato aroma compounds and to apply the
methods to various samples of tomatoes.   The third approach
involved the sensory evaluation of identified tomato volatiles to
determine their probable importance to fresh tomato aroma.

Qualitative Approach .   Aqueous condensate was obtained from
commercial tomato processors from the vacuum (ca. 100mm)
concentration of fresh tomato to give tomato paste.   The volatile
components from this condensate were obtained by continuous liquid
liquid extraction using first pentane and then diethyl ether.
Enough condensate was processed to give several grams of volatile
tomato concentrate.   The volatiles obtained from this extraction

showed little evidence of thermally produced Maillard type volatiles and even showed relatively high concentrations of (Z)-3-hexenal which the authors found difficult to isolate quantitatively by their method of vacuum steam distillation in the laboratory.   Comparison to volatiles isolated from fresh tomatoes in the laboratory, both by vacuum steam distillation-continuous extraction and by Tenax trapping, showed that they were similar qualitatively although there was considerable quantitative differences.

The volatile concentrate from the commercial condensate was first separated into 2 main fractions by micro distillation under reduced pressure (0.1mm Hg).   The distillation fractions were then resolved into their components by packed column GLC separation first with a 10 m Silicone SF96 packed column with further GLC resolution of the Silicone GLC fractions using a 3 m Carbowax 20-M column.   Infrared absorption spectra were measured with the separated components.   In some cases HNMR spectra and chemical ionization (C.I.) mass spectra were obtained.   This additional spectral data was particularly useful for the identification of some unusual compounds, i.e.,β-damascenone, 1-nitro-2-phenylethane, 1-nitro-3-methylbutane, β-cyclocitral and epoxy-β-ionone.

The nitro compounds, 1-nitro-3-methylbutane and 1-nitro-2-phenylethane, were particularly difficult to identify because they give very weak parent ions with electron ionization (E.I.).   However, C.I. mass spectra gave adequate M+1 ions. Figures 1 and 2 show the mass spectra (E.I.) of these compounds.   High resolution mass spectra also gave their empirical formula.   They were readily synthesized by reaction of 3-methylbutyl bromide or phenylethylbromide with sodium nitrite. The identification of 1-nitro-3-methylbutane in tomato had been reported previously by Wobben et al. (5) although they had not published any GLC or mass spectral data.   None of the other numerous studies of tomato volatiles (cf. 3) had reported finding this compound.   It is a relatively prominent component of fresh tomato occurring at a concentration as much as 200 ppb in some varieties such as Ace and related varieties but in other varieties it occurs at lower levels (10-50 ppb).   However, it does not seem to be important to fresh tomato aroma because it is a relatively weak odorant with an odor threshold of 150 ppb.

1-Nitro-3-methylbutane bears an interesting relationship to two other known volatile components of fresh tomatoes, isobutyl cyanide and 2-isobutylthiazole and to the non volatile amino acid leucine.   This relationship is shown in Figure 3.   They all show a similar skeletal arangement of carbon atoms to the left of the nitrogen.

1-Nitro-2-phenylethane bears a similar relationship to phenylalanine and phenylacetonitrile.   The nitro compounds are possibly formed by oxidation of these amino acids.   Stone et al., (6) had previously presented evidence (using radioactive isotopes) that leucine was one of the precursors of 2-isobutylthiazole. 1-Nitro-2-phenylethane is a moderately potent odorant with an odor threshold of 2 ppb and as later discussed probably contributes to the tomato aroma.

β-Damascenone had not been previously reported in tomatoes until

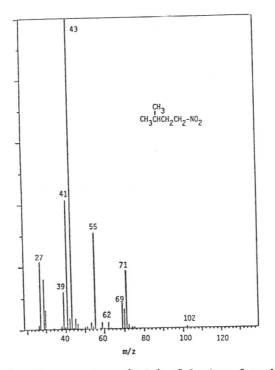

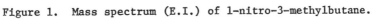

Figure 1. Mass spectrum (E.I.) of 1-nitro-3-methylbutane.

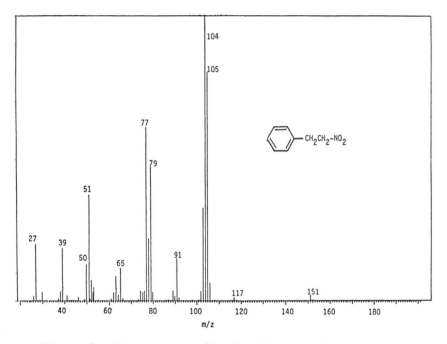

Figure 2.   Mass spectrum (E.I.) of 1-nitro-2-phenylethane.

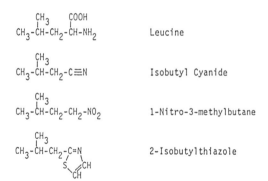

Figure 3.   Relationship of the structures of 1-nitro-3-methyl-butane, isobutyl cyanide and 2-isobutylthiazole to each other and to the amino acid leucine.

recently by some of the authors (7).   Its presence had been
well established in a number of other products such as apples (8)
and grapes (9).   It was identified by the authors in tomatoes by
comparison of its mass spectrum, capillary GLC retention data and
infrared absorption spectrum with those of an authentic sample.   In
reviewing mass spectral data on tomato volatiles they had recorded
in 1970 (10) some of the authors found that they had obtained a
mass spectrum of β-damascenone then but had not been able to
identify it.   In that year the structure of β-damascenone had
only just been elucidated and proved by synthesis by Demole et al.,
(11).   β-Damascenone has been shown to be a potent odorant (9,12)
which was verified by the authors who determined its odor threshold
to be 0.002 parts per billion (ppb) of water.   Its concentration
in fresh tomatoes was found to be 1-3 ppb (500-1500 times its
threshold) so that it seems very likely that β-damascenone is an
important aroma component of fresh tomatoes.

β-Cyclocitral had not been previously reported in tomatoes except
by the authors (2).   Structurally related to β-ionone it probably
also results from oxidative degradation of β-carotene.   With an
odor threshold of 5 ppb it is a moderately potent odorant.
However, its concentration in blended tomatoes is usually below
this concentration and it seems unlikely that it can make a
significant contribution to fresh tomato aroma.

Epoxy-β-ionone had been reported previously by Viani et al.,(13),
Schreier et al., (14) and Wobben et al., (5).   In the present
study besides the mass spectrum an infrared absorption spectrum was
also obtained and was found to be identical to that of an authentic
sample.   An odor threshold was determined in water solution to
be 100 ppb.   It is, therefore, a relatively weak odorant and as
its concentration, in all fresh tomato samples examined, is well
below this figure it seems unlikely that it can contribute to fresh
tomato aroma.

Quantitative Studies.   A method for the quantitative analysis of
fresh tomato volatiles using a saturated $CaCl_2$ solution to
deactivate tomato enzymes and Tenax trapping had been developed by
the authors (1,2).   The method also included the use of the
internal standards 3-pentanone, 2-octanone and anethole added (as a
dilute solution in water) just after the addition of the saturated
$CaCl_2$.   The authors had shown (2) that for a 64% recovery of a
compound a simple equation
$$Va = Vw / K$$
(where Va = volume of sweep gas, Vw = volume of aqueous solution
and K = air/water partition coefficient) gave the amount of sweep
gas needed for Tenax trapping.   There are some other factors
involved also, of course, such as adsorption of volatiles on the
glass walls of the flask and sweep gas outlet head.   From some
studies that the authors have carried out, this adsorption is
probably negligible for most compounds with boiling points less
than about 2-octanone but become more important with higher boiling
compounds such as eugenol.   As with adsorption on Tenax,
adsorbed compounds are also desorbed from the glass surface by the
continual flow of sweep gas over the surface.   The greater the
amount of sweep gas passed over the surface, the smaller will
be the amount of compound remaining on the glass surface.

The 1 hour trapping period is practical for typical quantitative studies where many samples are involved.   It would be desirable to shorten this time even more.   As can be seen from the above equation this might be accomplished by decreasing the volume of sample Vw which in turn decreases the required volume of air, Va.   Thus halving Vw halves Va (this ignores glass surface adsorption effects) to give a 1/2 hour trapping period. However, a smaller amount of sample would be available for GLC analysis which makes handling more difficult and sensitivity borderline for the lowest concentration components (such as β-ionone).

Studies with model systems of standard solutions of components in water using the same isolation procedure outlined by the authors for the tomato (2) showed satisfactory recoveries of most tomato volatiles .   A few compounds gave unacceptable recoveries for the 1 hour sweeping period used.   These were 2-phenylethanol which gave a 3% recovery (relative to anethole) and eugenol which gave a 0.5% recovery (relative to anethole) for the 1 hour sweep period . With 24 hour sweep periods both of these compounds gave better than 50% recoveries.

Experiments were also carried to determine the recovery of known amounts of tomato compounds added to samples of macerated tomato that had been previously heated to boiling and which contained very little of the fresh volatiles tested.   The recovery obtained from the macerated tomato relative to the internal standards was (within ca. ±10%) the same as that obtained for water solutions.

Comparison of the concentrations of the different components of the tomato was made previously (2) for the separated main parts of the tomato i.e. the skin, pulp, fluid and seed.   These studies had shown that the skin and pulp contained the highest concentrations of volatile components and that the seeds contained essentially none.

Concentrations found for a vine ripe (macerated) common commercial tomato line (GS-12) are shown in the first column of Table I. The data are the average of figures from three separate isolations.

Sensory Approach. The main sensory studies applied have been in the determination of odor thresholds of components and the calculation of odor unit values (Uo), the ratio of the concentration of the component in the food to its odor threshold in water.   The results of these studies are summarized in Table I.   It can be seen that (Z)-3-hexanal shows the most odor units followed by β-ionone, hexanal, β-damascenone, 1-penten-3-one, 3-methylbutanal, (E)-2-hexenal, 2-isobutylthiazole, 1-nitro-2-phenylethanol and (E)-2-heptenal.   β-Damascenone and 1-nitro-2-phenylethane are new members of this group.

Odor descriptions of dilute water solutions of β-damascenone and 1-nitro-2-phenylethane were also obtained using a panel of 18-20 judges. β-Damascenone as a 10 ppb solution in water was described as having an odor most similar to (1) prunes (2) apple (3) sweet character and (4) tomato in that order (i.e. being most like prunes).   The odor of 1 ppm solutions of 1-nitro-2-phenylethane were described as (1) green (2) geranium (3) tomato and (4) oily in that order.

Table I. Concentrations of major volatile fresh tomato
components using blending procedure, odor thresholds
in water solution and Log Odor Units. Compounds
listed in descending order of their Log odor units

Compound	Conc. ppb	Odor Thresh. ppb in H_2O	Log. Odor Units
(Z)-3-hexenal	12000	0.25	4.7
β-ionone	4	0.007	2.8
hexanal	3100	4.5	2.8
β-damascenone	1	0.002	2.7
1-penten-3-one	520	1	2.7
3-methylbutanal	27	0.2	2.1
(E)-2-hexenal	270	17	1.2
2-isobutylthiazole	36	3.5	1.0
1-nitro-2-phenylethane	17	2	0.9
(E)-2-heptenal	60	13	0.7
phenylacetaldehyde	15	4	0.6
6-methyl-5-hepten-2-one	130	50	0.4
(Z)-3-hexenol	150	70	0.3
2-phenylethanol	1900	1000	0.3
3-methylbutanol	380	250	0.2
methyl salicylate	48	40	0.08
geranylacetone	57	60	-0.02
β-cyclocitral	3	5	-0.2
1-nitro-3-methylbutane	59	150	-0.4
geranial	12	32	-0.4
linalool	2	6	-0.5
1-penten-3-ol	110	400	-0.6
(E)-2-pentenal	140	1500	-1.0
neral	2	30	-1.2
pentanol	120	4000	-1.5
pseudoionone	10	800	-1.9
isobutyl cyanide	13	1000	-1.9
hexanol	7	500	-1.9
epoxy-β-ionone	1	100	-2.0

ACKNOWLEDGMENTS

The authors thank Dr. M. Allen Stevens and Kevin Scott of Campbell
Institute for Rresearch and Technology, Davis, CA ; Dr. Rick
Falkenberg of Ragu Foods, Inc, Shelton CT and Mark Kimmel of
Stanislaus Food Products Co, Modesto, CA for samples of tomatoes
and tomato products and helpfull discussion.   We also wish to
thank Jean G. Turnbaugh for odor threshold determinations, Roger
England for CI and high resolution mass spectra and Mabry E. Benson
for NMR spectra (all of Western Regional Research Center).

LITERATURE CITED

1. Buttery, R. G. ; Teranishi, R.; Ling, L. C. "Fresh Tomato Aroma
Volatiles: A Quantitative Study". J. Agric. Food Chem. 1987, 35,
540.
2. Buttery, R. G.; Teranishi, R.; Ling, L. C.; Flath, R. A.;
Stern, D. J. " Quantitative Studies on Origins of Fresh Tomato
Aroma Volatiles". J. Agric. Food Chem. in press 1988.
3. Petro-Turza, M. "Flavor of Tomato and Tomato Products". Food
Reviews International. 1986-87, 2, 309.
4. Kornblum, N.; Larson, H. O.; Blackwood, R. K.; Mooberry, D. D;
Oliveto, E. P.; Graham, G. E. " A New Method for the Synthesis of
Aliphatic Nitro Compounds". J. Amer. Chem. Soc. 1956, 78, 1497.
5. Wobben, H. J.; de Valois, P. J.; ter Heid, R.; Boelens, H.;
Timmer, R. "Investigations Into the Composition of a Tomato
Flavour". Proc. IV, Int. Congress Food Sci. Technol.,Vol. I, 1974,
p.22.
6. Stone, E. J., Hall, R. M. and Kazeniac, S. J. "Precursors of
Tomato Volatiles". Paper presented 162nd National Meeting, American
Chemical Society, Washington, D.C. 1971.
7. Buttery, R. G.; Teranishi, R.; Ling, L. C. "Identification of
Damascenone in Tomato Volatiles". Chem. and Ind.(London) 1988,
238.
8. Nursten, H. E.; Woolfe, M. L. "Identification of Apple
Volatiles" J. Sci. Fd Agric. 1972, 23, 803.
9. Acree, T. E.; Braell, P.; Butts, R. M. "The Presence of
Damascenone in Cultivars of Vitis vinifera, rotundifolia and
labruscana". J. Agric. Food Chem. 1981, 29, 688.
10. Buttery, R. G.; Seifert, R. M.; Ling, L.C. Unpublished work,
1970.
11. Demole, E.; Enggist, P; Sauberli, U.; Stoll, M.; Kovats,E.
"Structure et synthese de la damascenone". Helv. Chim. Acta, 1970,
53, 541.
12. Pickenhagen, W., Firmenich SA, Geneva, Switzerland, personal
communication, 1987..
13. Viani, R.; Bricout, J.; Marion, J. P.; Muggler-Chaven, F.;
Reymond, D.; Egli, R. H. "Sur la Composition de l'Aroma de Tomate".
Helv. Chim. Acta 1969, 52, 887.
14. Schreier, P.; Drawert, F.; Junker, A. "Uber die quantitative
Zusammensetzung naturlicher und technologisch veranderter
pflanzlicher Aromen". Z. Lebensm. Unters.-Forsch 1977, 165, 23.

RECEIVED August 1, 1988

Chapter 18

Volatile Constituents of Pineapple (*Ananas Comosus* [L.] Merr.)

G. Takeoka[1], Ron G. Buttery[1], Robert A. Flath[1], Roy Teranishi[1], E. L. Wheeler[2], R. L. Wieczorek[2], and M. Guentert[3]

[1]Western Regional Research Center, Agricultural Research Service, U.S. Department of Agriculture, Albany, CA 94710
[2]Nabisco Brands, Inc., Schaeberle Technical Center, P.O. Box 1943, East Hanover, NJ 07936–1943
[3]Haarmann & Reimer GmbH, Postfach 1253, D–3450 Holzminden, Federal Republic of Germany

The volatiles of fresh pineapple (*Ananas comosus [L.] Merr.*) crown , pulp and intact fruit were studied by capillary gas chromatography and capillary gas chromatography-mass spectrometry. The fruit was sampled using dynamic headspace sampling and vacuum steam distillation-extraction. Analyses showed that the crown contains C_6 aldehydes and alcohols while the pulp and intact fruit are characterized by a diverse assortment of esters, hydrocarbons, alcohols and carbonyl compounds. Odor unit values, calculated from odor threshold and concentration data, indicate that the following compounds are important contributors to fresh pineapple aroma: 2,5-dimethyl-4-hydroxy-3(2H)-furanone, methyl 2-methylbutanoate, ethyl 2-methylbutanoate, ethyl acetate, ethyl hexanoate, ethyl butanoate, ethyl 2-methylpropanoate, methyl hexanoate and methyl butanoate.

Pineapple flavor has been the subject of extensive studies (1). Early work has been discussed in depth in a review by Flath (2). Pickenhagen et al. (3) reported the amounts of 2,5-dimethyl-4-hydroxy-3(2H)-furanone (furaneol) and its corresponding methyl ether in pineapples. With the aid of GC-sniffing, 2-propenyl hexanoate was identified in pineapple (4). Though the compound possesses a pineapple-like odor (5) it does not contribute to pineapple flavor at its naturally occurring levels in the fruit (6). In their study of the non-polar fraction Berger et al. (7) identified the sesquiterpenes α-copaene, β-ylangene, α-patchoulene (tentative), γ-gurjunene, germacrene D, α-muurolene and δ-cadinene. However, none of the compounds identified were responsible for the balsamic, fruity odor of the fraction. Nineteen new compounds including 1-(E,Z)-3,5-undecatriene and 1-(E,Z,Z)-3,5,8-undecatetraene were identified by Berger et al. (8). Due to their low odor thresholds the two unsaturated hydrocarbons may be important contributors to pineapple flavor. Ohta et al. (9) studied canned Philippine pineapple juice and reported methyl 4-acetoxyhexanoate and various carboxylic acids. The enantiomeric composition of various lactones, hydroxy and acetoxy esters occurring in pineapples has been reported by Tressl and co-workers (10,11).

Though pineapple flavor has been extensively studied relatively little work has been done on the odor properties and significance of the various constituents (6,8). This study investigates the odor contribution of various constituents. The volatiles from three parts of fresh pineapple, the crown, pulp and intact fruit were examined.

0097–6156/89/0388–0223$06.00/0

EXPERIMENTAL SECTION

Materials. Fresh pineapples (*Ananas comosus (L.) Merr.*) var. Smooth Cayenne were obtained via air freight from Hawaii.

Dynamic Headspace Sampling. The crown, intact fruit and blended pulp volatiles were isolated using dynamic headspace sampling.

Four crowns (total weight, 765g) were placed in a 9L Pyrex glass container (a modified desiccator). A Pyrex head to allow the passage of air into and out was fitted into a standard ground glass joint in the upper part of the container. Purified air (passed through activated carbon) was passed over the leaves at a flow rate of 3L/min. The air exiting the desiccator was passed through a Tenax trap. The traps were constructed of Pyrex glass tubing and terminated in standard ball and socket joints. The traps were loaded with 10g of 60/80 mesh Tenax GC (Alltech Associates, Deerfield, IL) producing a column 14 cm long X 2.2 cm i.d. The trapping was continued for 24 h at room temperature (ca. 24 °C). The collected volatiles were eluted from the Tenax trap with 80 mL of freshly distilled diethyl ether containing ca. 0.001% Ethyl antioxidant 330 (1,3,5-trimethyl-2,4,6-tris [3,5-di-tert-butyl-4-hydroxybenzyl]benzene). The ether extract was carefully concentrated with a Vigreux column to a final volume of ca. 100 µl.

Two intact pineapples (without crowns, total weight, 3.2 kg) were placed into a 9L glass container. The volatiles were collected, eluted and concentrated in the same manner as described above.

The pineapple pulp was sampled in the following manner. The skin was removed from the fruit. The flesh (400g) was cut into chunks and placed in a Waring blender with 200 mL of saturated $CaCl_2$ solution. The mixture was blended for 30 s. Fifteen milliliters of a water solution containing 20 ppm 3-heptanone and fifteen milliliters of a water solution containing 20 ppm 6-methyl-5-hepten-2-one was then added and the mixture was blended for 15 s. The mixture was placed in a 2L round-bottomed flask. Two hundred mL of water was added to the flask. Purified air (3L/min) was passed over the surface of the vigorously stirred mixture via a Teflon tube and exited out of the flask through a Tenax trap of the same dimensions as described above. The sampling apparatus is shown in Figure 1. The mixture was sampled for 3 h. The trapped material was eluted and the extract was concentrated as described above.

Vacuum Steam Distillation-Extraction. Pineapple pulp (1.0 kg) was blended with 1 L water for 20 s in a Waring blender. Three batches were prepared using a total of 3.13 kg fruit pulp. The blended material was added to a 12 L round-bottomed flask. Sixty milliliters of antifoam solution was added. The antifoam solution was prepared by adding 12 mL of Hartwick antifoam 50 emulsion to 900 mL of water in a 1 L flask and boiling until the volume was reduced to 600 mL. A modified Likens-Nickerson steam distillation extraction head (12) was employed. A 250 mL round-bottomed flask containing 125 mL hexane (containing 0.001% of Ethyl antioxidant 330) was attached to the solvent arm of the extraction head. Simultaneous steam distillation-extraction (SDE) under 60mm Hg was continued for 3 h. The resulting hexane extract was chilled to -20 °C to freeze out residual water. The extract was quickly decanted and then concentrated under 60mm Hg with a Vigreux column to a final volume of ca. 500 µl.

Gas Chromatography. A Hewlett-Packard 5890 gas chromatograph (Hewlett-Packard, Avondale, PA) with a flame ionization detector (FID), equipped with a 60 m X 0.32 mm i.d. DB-WAX column (d_f = 0.25 µm, bonded polyethylene glycol, J&W Scientific, Folsom, CA) was employed. Helium carrier gas was used at a flow rate of 1.64 mL/min (30°C). The oven temperature was programmed from 30°C (4 min isothermal) to 180°C at 2°C/min. A split ratio of 1:28 was used. The injector and detector were maintained at 200°C and 220°C, respectively. A 60 m X 0.32 mm i.d. DB-1 column (d_f = 0.25 µm, bonded dimethyl polysiloxane, J&W Scientific, Folsom, CA) was used to analyze the sample prepared by vacuum steam distillation-extraction. Helium carrier gas was used a t a flow rate of 1.60 mL/min (30 °C). The oven temperature was programmed from 30 °C (4 min isothermal) to

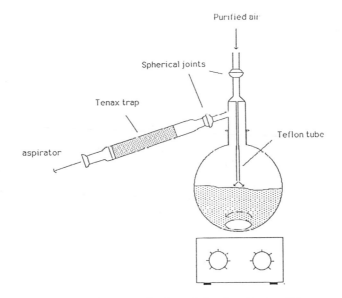

Figure 1.   Apparatus for dynamic headspace sampling.

210 °C at 2 °C/min. A split ratio of 1:22 was used. Data processing was performed with an HP 5895 GC ChemStation.

<u>Gas Chromatography-Mass Spectrometry.</u> A Finnigan MAT 4500 GC/MS/INCOS system (Finnigan MAT, San Jose, CA) was used for analysis. A 60 m X 0.32 mm i.d. DB-WAX column was employed for the headspace samples. The oven temperature was programmed from 50°C to 180°C at 2°C/min. A 60 m X 0.32 mm i.d. DB-1 column was used for the sample prepared by vacuum SDE. The oven temperature was programmed from 30 °C (4 min isothermal) to 210 °C at 2 °C/min. Helium was used as the carrier gas at a flow rate of 3.2 mL/min. The column outlet was inserted directly into the ion source block. The instrument was operated in the electron impact mode at an ionization voltage of 70 eV. The mass spectrometer was repetitively scanned from 33 to 350 m/z in a one-second cycle.

<u>Reference Compounds.</u> Authentic reference compounds were obtained from commercial sources, synthesized or isolated from essential oils. Methyl 3-acetoxybutanoate was prepared by acetylation of methyl 3-hydroxybutanoate with acetic anhydride. Methyl 3-acetoxybutanoate had the following mass spectrum, m/z (relative intensity): 117(27), 100(23), 87(4), 85(11), 74(2), 69(37), 59(20), 43(100). Methyl (Z)-3-hexenoate was prepared from methyl (E)-3-hexenoate by reaction with the equilibration catalyst, benzenethiol (13). 3-Methylthiopropyl acetate was formed by acetylation of 3-methylthiopropanol with acetic anhydride. 3-Methylthiopropyl acetate had the following mass spectrum, m/z (relative intensity): 148 (43), 105(5), 101(3), 90(5), 89(5), 88(100), 75(19), 73(70), 61(44), 47(12), 45(11), 43(92). Methyl 4-acetoxyhexanoate was prepared by treatment of γ-hexalactone with base followed by acetylation with acetic anhydride. Methyl 4-acetoxyhexanoate had the following mass spectrum, m/z (relative intensity): 159(5), 145(8), 128(8), 117(51), 115(40), 113(21), 101(5), 97(7), 88(10), 85(28), 74(8), 69(11), 55(8), 43(100). Methyl 4-acetoxyoctanoate was prepared by treatment of γ-octalactone with base followed by acetylation with acetic anhydride. Methyl 4-acetoxyoctanoate had the following mass spectrum, m/z (relative intensity): 173(13), 159(14), 143(29), 141(40), 129(11), 124(13), 117(100), 115(18), 101(15), 88(16), 85(36), 74(13), 55(17), 43(99).

<u>Odor Thresholds.</u> These were determined on GLC purified samples using methods previously described (14), with a panel of 16-20 judges.

RESULTS AND DISCUSSION

The samples were analyzed by capillary gas chromatography and gas chromatography/mass spectrometry (GC/MS). Identifications were verified by comparing the component's mass spectrum and experimental retention index (I) with that of an authentic reference standard. The retention system proposed by Kovats (15) was utilized. When standards were not available the identifications were considered tentative.

The volatiles of the crown, pulp and whole intact fruit were examined by dynamic headspace sampling using a procedure developed in our laboratory (16). The method uses a fast flow of sweeping gas (3L/min) onto large Tenax traps.

Table I lists the compounds identified in a headspace sample of pineapple crowns. The sample was characterized by low levels of very few volatiles. The C_6 compounds, hexanal, (Z)-3-hexenal, (E)-2-hexenal and (Z)-3-hexenol were probably produced enzymatically in response to tissue damage from cutting (17). Hydrocarbons identified include styrene, and the monoterpenes, α- and β-pinene.

A GC/FID chromatogram of pulp headspace volatiles is shown in Figure 2. Components identified in the pulp along with their concentrations are listed in Table II. Quantitation was based on two internal standards, 3-heptanone and 6-methyl-5-hepten-2-one. These compounds were chosen for their chemical stability and their elution in relatively uncrowded regions of the chromatogram. Though earlier studies (16) showed good solute recovery using this sampling technique the reported concentrations should be considered as approximate values since solute recoveries and flame ionization detector

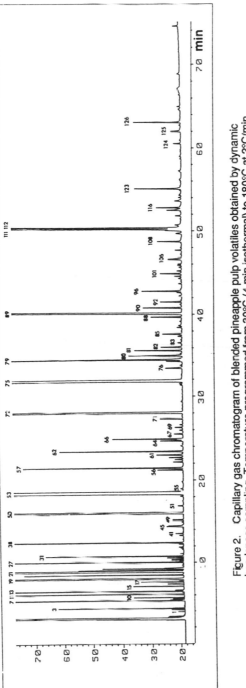

Figure 2. Capillary gas chromatogram of blended pineapple pulp volatiles obtained by dynamic headspace sampling. Temperature programmed from 30°C (4 min isothermal) to 180°C at 2°C/min on a 60m X 0.32 mm i.d. DB-WAX column. The peak numbers correspond to the numbers in Table II.

(FID) response factors were not determined for each identified pineapple constituent. Qualitative and quantitative differences were noted between samples and it was observed (by GC/MS) that ethyl heptanoate eluted as a shoulder on the 6-methyl-5-hepten-2-one peak in some runs. Therefore, the use of 6-methyl-5-hepten-2-one was discontinued in later runs. The major compounds found were methyl hexanoate, methyl 2-methylbutanoate, methyl butanoate, methyl octanoate, methyl 3-methylthiopropanaote and ethyl acetate.

Table I. Pineapple Crown Volatiles

constituent	DBWAX exp.	ref.
ethyl acetate	889	890
ethanol	940	927
pentanal		975
α-pinene	1015	1012
hexanal	1081	1078
β-pinene	1096	1093
(Z)-3-hexenal	1145	1138
(E)-2-hexenal	1213	1215
styrene	1253	1252
(Z)-3-hexenol	1382	1378

To assess the relative contribution of the identified constituents to the total odor the number of odor units (U_o) was calculated. Guadagni et al. (14) defined the odor unit as the ratio of the concentration of the compound and its odor threshold. Table III lists the odor units of some pineapple constituents calculated from their concentrations and odor thresholds. Compounds are listed in decreasing order of their number of odor units. The list is dominated by esters. It is likely that ethyl acetate makes a larger contribution to pineapple flavor than is shown on the table. Preliminary experiments have indicated that the % recovery of ethyl acetate with this headspace sampling procedure is very low (approximately 5 %), presumably due to breakthrough on the Tenax trap. Methyl 3-acetoxyhexanoate, methyl (E)-3-octenoate and ethanol are present in concentrations less than their odor thresholds and hence are expected to make little or no contribution to the pineapple odor. The odor thresholds for the unsaturated hydrocarbons, 1,3,5-undecatriene and 1,3,5,8-undecatetraene have not been determined and therefore they are not included on the table though they probably make an important contribution to pineapple flavor. The potent odor character of 1-(E,Z)-3,5-undecatriene and 1-(E,Z,Z)-3,5,8-undecatetraene has been described by Berger et al. (8). The configuration of the double bond in the 5 position is crucial; the corresponding isomers, 1-(E,E)-3,5-undecatriene and 1-(E,E,Z)-3,5,8-undecatetraene have odor thresholds 10^6 and 10^4 times higher, respectively (8). Another important volatile not included in Table III is 2,5-dimethyl-4-hydroxy-3(2H)-furanone (furaneol). This highly polar constituent does not steam distill and was not recovered using our sampling techniques. Its contribution to pineapple flavor was estimated using values published in the literature. Pickenhagen et al. (3) found 7.4 ppm of furaneol in pineapples from the Ivory Coast. This potent odorant has an odor threshold of 0.03 ppb (18). Thus the odor unit value calculated for furaneol is approximately 2.5×10^5. Therefore, it must be one of the major contributors to pineapple flavor. It is interesting to examine the odor descriptions of the compounds with the highest odor units (Table IV). These esters largely possess pineapple or apple odors. Ethyl 2-methylbutanoate has been found to be an important character impact compound in Delicious apple essence (19).

Constituents identified in the intact ripe pineapple headspace sample are listed in Table V. Figure 3 shows a GC/FID chromatogram of pineapple headspace. The ester fraction comprises about 81% of the total area. Quantitatively, the major constituents were methyl

Table II. Volatile Constituents of Pineapple - Blended Pulp (Headspace)

peak no.[a]	constituent	DBWAX exp.	ref.	conc. ppb
7	ethyl acetate	887	886	503
10	methyl propanoate	904	904	14
11	methyl 2-methylpropanoate	922	920	64
13	ethanol	936	937	192
15	ethyl propanoate	957	952	17
16	ethyl 2-methylpropanoate	966	962	6
17	propyl acetate	974	974	17
19	methyl butanoate	984	982	2026
21	methyl 2-methylbutanoate	1007	1009	2079
22	2-methylpropyl acetate	1011	1010	41
23	chloroform	1016	1015	35
24	methyl 3-methylbutanoate	1017	1018	9
27	ethyl butanoate	1035	1035	92
31	ethyl 2-methylbutanoate	1051	1048	66
38	methyl pentanoate	1081	1083	150
46	3-methylbutyl acetate	1118	1120	8
49	ethyl pentanoate	1131	1132	6
50	3-heptanone (internal standard)			
53	methyl hexanoate	1182	1180	3442
56	(methyl 5-hexenoate)[b,c]	1224		15
57	ethyl hexanoate	1228	1228	99
61	(methyl (Z)-3-hexenoate)[b,c]	1252		22
62	methyl (E)-3-hexenoate	1258	1259	73
66	methyl heptanoate	1279	1279	48
72	6-methyl-5-hepten-2-one (internal standard)			
75	methyl octanoate	1380	1380	1451
78	methyl 2,4-hexadienoate[c]	1414	1447	11
79	(methyl (Z)-4-octenoate)[b]	1417		133
80	ethyl octanoate	1426	1426	37
81	1,3,5,8-undecatetraene+ acetic acid	1431		31
82	methyl (E)-3-octenoate	1446	1446	15
85	α-copaene	1469	1471	15
88	dimethyl malonate	1499	1500	19
89	methyl 3-methylthiopropanoate	1506	1505	596
92	methyl 3-acetoxybutanoate	1529	1530	14
96	ethyl 3-methylthiopropanoate	1551	1551	27
106	methyl (Z)-4-decenoate	1612	1623	14
112	methyl 3-acetoxyhexanoate	1676	1678	166
115	ethyl 3-acetoxyhexanoate	1712	1712	13
123	(methyl 5-acetoxyhexanoate)[b]	1759		33
126	methyl 5-acetoxyoctanoate	1904	1904	34

[a]the peak numbers correspond to the numbers in Figure 2. [b]tentative identifications enclosed in parentheses. [c]identified for the first time in pineapple.

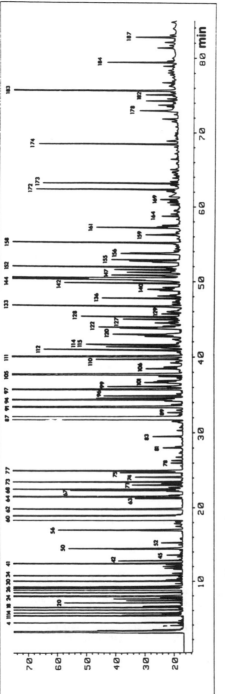

Figure 3.   Capillary gas chromatogram of intact pineapple headspace volatiles. Temperature programmed from 30°C (4 min isothermal) to 180°C at 2°C/min on a 60m X 0.32 mm i.d. DB-WAX column. The peak numbers correspond to the numbers in Table V.

hexanoate, methyl 2-methylbutanoate, methyl octanoate, ethyl acetate and methyl butanoate. These esters which make up the bulk of the intact fruit headspace sample (>69%) were also major compounds found in the pulp sample. The presence of a variety of unsaturated hydrocarbons is noteworthy. This fraction accounts for about 8.6% of the

Table III. Odor Thresholds and Odor Units for Selected Constituents in Blended Hawaiian Pineapples (Pulp)

Constituent	Odor Threshold ppb	Odor Units $(U_o)^a$
methyl 2-methylbutanoate	0.25	8316
ethyl 2-methylbutanoate	0.3	220
ethyl acetate	5	100.6
ethyl hexanoate	1	99
ethyl butanoate	1	92
ethyl 2-methylpropanoate	0.1	60
methyl hexanoate	70	49
methyl butanoate	60	33.8
methyl heptanoate	4	12
methyl 2-methylpropanoate	7	9.1
methyl pentanoate	20	7.5
methyl octanoate	200	7.3
methyl (Z)-4-decenoate	3	4.7
ethyl pentanoate	1.5	4
3-methylbutyl acetate	2	4
ethyl 3-methylthiopropanoate	7	3.9
methyl 3-methylthiopropanaote	180	3.3
ethyl propanoate	10	1.7
methyl 3-acetoxyhexanoate	190	0.9
methyl (E)-3-octenoate	150	0.1
ethanol	100000	0.002

$^a U_o$ = concentration of the compound divided by its threshold concentration.

Table IV. Odor Descriptions and Odor Units for Selected Constituents in Blended Hawaiian Pineapples (Pulp)

Constituent	Odor Units	Odor Description (Fenaroli (5))
methyl 2-methylbutanoate	8316	pungent, fruity
ethyl 2-methylbutanoate	220	apple (19)
ethyl acetate	100.6	ether-like reminiscent of pineapple
ethyl hexanoate	99	powerful fruity with pineapple-banana note
ethyl butanoate	92	fruity with pineapple undernote
ethyl 2-methylpropanaote	60	apple like
methyl hexanoate	49	ether-like reminiscent of pineapple
methyl butanoate	33.8	apple-like
methyl heptanoate	12	strong, almost fruity, orris-like

Table V. Volatile Constituents of Pineapple - Whole Intact Fruit (Headspace)

peak no.[a]	constituent	exp.	ref.	%area[b]
		I DBWAX		
11	ethyl acetate	887	886	5.91
14	methyl propanoate	904	904	0.14
16	methyl 2-methylpropanoate	922	920	0.21
18	ethanol	936	927	1.56
20	ethyl propanoate	957	952	0.12
21	ethyl 2-methylpropanoate	966	962	0.06
22	propyl acetate	973	974	0.08
24	methyl butanoate	984	982	5.29
26	methyl 2-methylbutanoate	1008	1009	11.42
27	2-methylpropyl acetate	1011	1010	0.42
28	chloroform	1016	1015	0.32
29	methyl 3-methylbutanoate	1017	1016	0.09
30	ethyl butanoate	1035	1035	0.70
34	ethyl 2-methylbutanoate	1051	1048	0.63
36	ethyl 3-methylbutanoate	1067	1065	tr
37	butyl acetate[d]	1070	1068	0.03
40	hexanal	1078	1078	0.02
41	methyl pentanoate	1082	1083	0.82
42	2-methylpropanol	1088	1084	0.09
45	diethyl carbonate?	1099	1102	0.04
50	3-methylbutyl acetate	1118	1118	0.19
52	ethyl pentanoate	1131	1132	0.04
56	myrcene[d]	1160	1160	0.30
59	2-heptanone[d]	1176	1177	tr
60	methyl hexanoate	1187	1180	36.57
61	limonene[d]	1188	1183	0.22
62	3-methylbutanol	1200	1201	0.55
63	(methyl 5-hexenoate)[c,d]	1224		0.08
64	ethyl hexanoate	1228	1228	1.14
66	(monterpene)[c]	1239		0.03
67	(E)-β-ocimene[d]	1241	1243	0.20
68	styrene[d]	1243	1242	0.34
71	(methyl (Z)-3-hexenoate)[c,d]	1252		0.12
73	methyl (E)-3-hexenoate	1258	1259	0.31
74	acetoin	1267	1268	0.12
77	methyl heptanoate	1279	1279	0.76
81	(methyl 4-heptenoate)[c,d]	1323		0.04
83	hexanol[d]	1346	1348	0.05
87	methyl octanoate	1383	1380	10.27
88	1-(E,Z)-3,5-undecatriene	1384	1382	0.39
89	1-(E,E)-3,5-undecatriene	1391	1393	0.03
91	(sesquiterpene)[c]	1402		1.53
92	(sesquiterpene)[c]	1407		0.03
94	(methyl (Z)-4-octenoate)[c]	1418		0.89
95	(sesquiterpene)[c]	1421		0.17
96	ethyl octanoate	1426	1426	0.15
97	1,3,5,8-undecatetraene	1440		1.42
99	methyl (E)-3-octenoate	1446	1446	0.16
101	1,3,5,8-undecatetraene	1454		0.10
102	(sesquiterpene)[c]	1455		0.05

Table V. *Continued*

peak no.[a]	constituent	DBWAX exp.	ref.	%area[b]
104	methyl 3-hydroxybutanoate	1466	1467	0.05
105	α-copaene	1471	1471	1.60
106	methyl nonanoate	1481	1481	0.08
108	methyl (E)-2-octenoate[d]	1484	1485	0.05
110	dimethyl malonate	1499	1500	0.18
111	methyl 3-methylthiopropanoate	1506	1505	1.47
114	methyl 3-acetoxybutanoate	1529	1530	0.27
115	2,3-butanediol[d]	1529	1536	e
119	ethyl 3-methylthiopropanoate	1550	1551	0.08
120	(sesquiterpene)[c]	1552		0.19
122	β-copaene[d]	1567	1567	0.17
127	methyl decanoate	1583	1584	0.11
128	(sesquiterpene)[c]	1589		0.22
133	methyl (Z)-4-decenoate	1612	1623	0.42
136	methyl 3-hydroxyhexanoate	1630	1632	0.16
142	γ-muurolene[d]	1664	1665	0.27
145	methyl 3-acetoxyhexanaote	1676	1678	0.66
147	(sesquiterpene)[c]	1682		0.20
148	β-selinene[d]	1688	1689	0.16
150	α-selinene[d]	1694	1694	0.16
152	α-muurolene	1700	1700	0.46
153	ethyl 3-acetoxyhexanoate	1712	1712	0.05
154	(sesquiterpene)[c]	1714		0.12
156	δ-cadinene	1732	1734	0.15
158	(methyl 5-acetoxyhexanoate)[c]	1759		0.45
161	2-phenylethyl acetate[d]	1794	1794	0.18
172	2-phenylethanol	1889	1890	0.30
173	methyl 5-acetoxyoctanoate	1904	1904	0.27
183	(ethylphenol)[c,d]	2155		0.97

[a]the peak numbers correspond to the numbers in Figure 3. [b]peak area percentage of the total FID area excluding the solvent peaks (assuming all response factors of 1). "tr" represents less than 0.02%. [c]tentative identifications enclosed in parentheses. [d]identified for the first time in pineapple. [e]merged with previous peak.

Table VI. Volatile Constituents of Pineapple - Blended Pulp (Vacuum SDE)

peak no.	constituent	DB-1 exp.	DB-1 ref.	%area[a]
1	methyl butanoate	709	705	0.55
2	methyl cyclohexane[b]	719	718	0.14
4	dimethyl hexane[b]	735		0.21
5	dimethyl hexane[b]	737		0.08
7	2-methylpropyl acetate	764	764	0.04
9	methyl 2-methylbutanoate	768	768	1.03
10	diethyl carbonate	769	769	0.02
13	3-hexanol	784	784	0.01
14	ethyl butanoate	788	789	0.43
16	methyl pentanoate	810	810	0.24
18	ethyl 2-methylbutanoate	842	842	0.29
19	ethyl 3-methylbutanaote	844	845	0.01
21	3-methylbutyl acetate	866	866	0.17
22	2-methylbutyl acetate	868	869	0.08
23	ethyl pentanoate	888	888	0.10
24	(methyl 5-hexenoate)[c,d]	894		0.06
25	dimethyl malonate	899	897	0.06
27	3-methylbut-2-enyl acetate[d]		902	tr
28	methyl hexanoate	914	910	14.72
29	(methyl (Z)-3-hexenoate)[c,d]	917	916	0.05
30	ester	918		0.04
31	methyl (E)-3-hexenoate	921	920	0.51
33	methyl (E)-2-hexenoate	947	948	0.10
37	ethyl hexanoate	988	986	6.80
39	methyl 2,4-hexadienoate[d]	991	991	0.06
40	methyl 3-methylthiopropanaote	1000	992	17.41
41	γ-hexalactone	1003	1003	0.06
42	methyl heptanoate	1010	1009	0.20
43	methyl 3-acetoxybutanoate	1017	1016	0.35
45	methyl 3-hydroxyhexanoate	1029	1026	0.03
46	(Z)-β-ocimene[d]	1030	1026	0.02
47	2,5-dimethyl-4-methoxy-3(2H)-furanone	1032	1031	0.05
48	(E)-β-ocimene[d]	1041	1037	0.13
52	ethyl 3-methylthiopropanaote + ?	1075	1072	3.19
55	ethyl heptanaote	1084	1080	0.10
56	linalool?	1086	1083	0.04
58	3-methylthiopropyl acetate[d]	1091	1091	0.10
60	methyl (Z)-4-octenoate	1097	1097	0.77
62	ester	1101		0.03
63	ethyl 3-hydroxyhexanoate	1105	1103	0.05
64	methyl octanoate +	1111	1107	6.82
	methyl (E)-3-octenoate		1110	
68	methyl phenylacetate[d]	1145	1144	0.10
69	methyl (E)-2-octenaote[d]	1150	1150	0.03
71	4-terpineol	1158	1159	0.07
72	1-(E,Z)-3,5-undecatriene+	1165	1165	0.70
	1-(E,Z,Z)-3,5,8-undecatetraene			
73	α-terpineol	1169	1170	0.07
74	ethyl (Z)-4-octenoate?	1173		0.38
75	1-(E,E,Z)-3,5,8-undecatetraene	1175	1175	0.06
78	methyl 3-acetoxyhexanoate + ?	1186	1176	15.17

Table VI. *Continued*

peak no.	constituent	DB-1 exp.	ref.	%area[a]
79	methyl 4-acetoxyhexanoate	1205	1203	0.56
80	methyl nonanoate	1207	1207	0.13
82	(methyl 5-acetoxyhexanoate)[c]	1226		1.33
85	ethyl 3-acetoxyhexanoate	1252	1250	2.24
87	ethyl 4-acetoxyhexanoate?	1275		0.19
88	ethyl nonanoate	1282	1279	0.04
89	methyl (Z)-4-decenoate + ?	1291	1289	3.09
90	ethyl 5-acetoxyhexanoate	1293	1293	0.16
92	methyl decanaote	1308	1307	0.38
94	(sesquiterpene)[c]	1321		0.14
95	(sesquiterpene)[c]	1327		0.11
98	cyclocopacamphene[d]	1360	1361	0.13
99	(sesquiterpene)[c] + ?	1363		1.44
100	(methyl 3-acetoxyoctanoate)[c]	1365		0.28
101	α-copaene	1370	1370	1.72
103	methyl 4-acetoxyoctanoate + ethyl decanoate	1381	1379	0.80
104	methyl 5-acetoxyoctanoate	1389	1387	2.36
105	(α-gurjunene)[c,d]	1399	1401	0.19
107	β-copaene[d]	1418	1418	0.16
108	(sesquiterpene)[c]	1422		0.22
111	ethyl 3-acetoxyoctanoate?	1433		0.09
114	ethyl 4-acetoxyoctanoate?	1447		0.09
115	ethyl 5-acetoxyoctanoate	1457	1458	0.70
116	γ-muurolene[d] + (sesquiterpene)[c]	1465	1464	0.75
121	α-selinene[d] + (sesquiterpene)[c]	1482	1489	0.38
122	α-muurolene	1488	1487	1.51
123	(sesquiterpene)[c]	1489		0.64
127	(sesquiterpene)[c]	1501		0.08
128	δ-cadinene	1508	1513	0.51
131	calacorene?	1520		0.05
137	ethyl dodecanoate	1579	1578	0.06
146	γ-dodecalactone	1631	1635	0.04
151	geranyl hexanoate?	1731		0.04
156	ethyl tetradecanoate	1778	1778	0.02
160	ethyl hexadecanoate	1978	1978	0.02

[a]peak area percentage of the total FID area excluding the solvent peaks (assuming all response factors of 1). the values are approximate since there are known pineapple constituents co-eluting with the solvent peaks. "tr" represents less than 0.01%. [b]solvent contaminant. [c]tentative identifications enclosed in parentheses. [d]identified for the first time in pineapple.

total area. This sampling procedure is well suited for the analysis of unsaturated hydrocarbons which may undergo enzymatic and/or oxidative degradation with conventional sampling and separation techniques (20). Monoterpene hydrocarbons were identified for the first time in pineapple. These include myrcene, limonene and (E)-β-ocimene. None of these compounds were found in the pulp sample. Similarly, Takeoka et al. (21) identified a number of monoterpene hydrocarbons in intact nectarine headspace while none were found in a sample prepared by vacuum distillation followed by liquid-liquid extraction.

Previous studies (7) showed evidence for the presence of at least 20 sesquiterpene hydrocarbons in pineapple. These researchers identified 7 sesquiterpenes (1 tentative). We confirmed the presence of 3 of the sesquiterpenes and identified β-copaene, γ-muurolene, α- and β-selinene in addition. Mass spectral data indicated the presence of other sesquiterpene hydrocarbons and oxygenated sesquiterpenes which we have been unable to characterize at present.

Methyl (E)-2-octenoate was found for the first time in pineapple. It had been previously reported in pears (22) and soursop (23). Its formation from methyl (E)-3-octenoate (a pineapple constituent) by 2,3-(E,E)-enoyl-CoA-isomerase was postulated by Berger and Kollmannsberger (6)

The vacuum SDE method confirmed the presence of nearly all of the constituents identified using dynamic headspace sampling and revealed many additional compounds. The method was more effective in extracting the less volatile constituents such as long chain esters. In contrast to the previous runs this sample was chromatographed on a non-polar DB-1 column. The constituents identified in the pulp sample prepared by vacuum SDE are listed in Table VI. The %area values should be considered as only approximate since known pineapple constituents such as ethyl acetate, methyl propanoate, methyl 2-methylpropanoate, ethyl propanoate, ethanol, propyl acetate, and ethyl 2-methylpropanoate co-elute with the solvent peaks and hence could not be included in the quantitation.

The sulfur containing ester, 3-methylthiopropyl acetate, is reported for the first time in pineapple. This compound bears a relationship to the major esters, methyl and ethyl 3-methylthiopropanoate. It has been previously identified in apples (24), beer (25), wine (26) and whisky (27). The sesquiterpene hydrocarbon, cyclocopacamphene, is another newly reported pineapple constituent. This constituent was previously reported in vetiver oil (28).

The presence of the recently reported ester, methyl 4-acetoxyoctanoate (29) was confirmed.

ACKNOWLEDGMENT

The authors thank Prof. Dr. W. Boland, Universitat Karlsruhe, for supplying samples of 1-(E,Z,Z)-3,5,8-undecatetraene and 1-(E,E,Z)-3,5,8-undecatetraene.

Literature cited

1.   van Straten, S.; Maarse, H. (eds.) In "Volatile Compounds in Food - Qualitative Data", 5th ed.; Division for Nutrition and Food Research - TNO, Zeist, 1983.
2.   Flath, R.A. In "Tropical and Subtropical Fruits Composition, Properties and Uses"; Nagy, S.; Shaw, P., Eds.; AVI, Westport, 1980, p. 157.
3.   Pickenhagen, W.; Velluz, A.; Passerat, J.-P.; Ohloff, G. J. Sci. Food Agric. 1981, 32, 1132.
4.   Nitz, S.; Drawert, F. Chem. Mikrobiol. Technol. Lebensm. 1982, 7, 148.
5.   Fenaroli, G. In "Fenaroli's Handbook of Flavor Ingredients"; second edition; Furia, T.E.; Bellanca, N., Eds.; CRC Press, Cleveland, 1975.
6.   Berger, R.G.; Kollmannsberger, H. In "Topics In Flavour Research"; Berger, R.G.; Nitz, S.; Schreier, P., Eds.; H. Eichhorn, Marzling, 1985, p. 305-320.
7.   Berger, R.G.; Drawert, F.; Nitz, S. J. Agric. Food Chem. 1983, 31, 1237.
8.   Berger, R.G.; Drawert, F.; Kollmannsberger, H.; Nitz, S.; Schraufstetter, B. J. Agric. Food Chem. 1985, 33, 232.
9.   Ohta, H.; Kinjo, S.; Osajima, Y. J. Chromatogr. 1987, 409, 409.

10. Tressl, R.; Engel, K.-H.; Albrecht, W.; Bille-Abdullah, H. In "Characterization and Measurement of Flavor Compounds"; Bills, D.D.; Mussinan, C., Eds.; ACS Symposium Series vol. 289, Washington, D.C., 1985, p. 43-60.
11. Tressl, R.; Heidlas, J.; Albrecht, W.; Engel, K.-H. In "Bioflavour '87"; Schreier, P., Ed.; Walter de Gruyter & Co., Berlin, New York, 1988, p. 221-236.
12. Schultz, T.H.; Flath, R.A.; Mon, T.R.; Eggling, S.B.; Teranishi, R. J. Agric. Food Chem. 1977, 25, 446.
13. Henrick, C.A.; Willy, E.; Baum, J.W.; Baer, T.A.; Garcia, B.A.; Mastre, T.A.; Chang, S.M. J. Org. Chem. 1975, 40, 1.
14. Guadagni, D.G., Buttory, R.G.; Harris, J. J. Sci. Food Agric. 1966, 17, 142.
15. Kovats, E. Helv. Chim. Acta 1958, 41, 1915.
16. Buttery, R.G.; Teranishi, R.; Ling, L.C. J. Agric. Food Chem. 1987, 35, 540.
17. Schreier, P. In "Chromatographic Studies of Biogenesis of Plant Volatiles", Huthig, Heidelberg, 1984.
18. Honkanen, E.; Pyysalo, T.; Hirvi, T. Z. Lebensm. Unters. Forsch. 1980, 180.
19. Flath, R.A.; Black, D.R.; Guadagni, D.G.; McFadden, W.H.; Schultz, T.H. J. Agric. Food Chem. 1967, 15, 29.
20. Tressl, R.; Drawert, F.; Heimann, W.; Emberger, R. Phytochemistry 1970, 9, 2327.
21. Takeoka, G.R.; Flath, R.A.; Guentert, M.; Jennings, W. J. Agric. Food Chem. 1988, 36, 553.
22. Heinz, D.E.; Jennings, W.G. J. Food Sci. 1966, 31, 69.
23. MacLeod, A.J.; Pieris, N.M. J. Agric. Food Chem. 1981, 29, 488.
24. Schreier, P.; Drawert, F.; Mick, W. Lebensm.- Wiss. u. Technol. 1978, 11, 116.
25. Schreier, P.; Drawert, F.; Junker, A. Brauwissenschaft 1974, 27, 205.
26. Schreier, P.; Drawert, F.; Junker, A. Z. Lebensm. Unters. Forsch. 1974, 154, 279.
27. Masuda, M.; Nishimura, K. J. Food Sci. 1982, 47, 101.
28. Andersen, N.H. Tetrahedron Lett. 1970, 4651.
29. Engel, K.-H.; Heidlas, J.; Albrecht, W.; Tressl, R. Paper presented at American Chemical Society Meeting, 5-10 June 1988; Toronto, Canada.

RECEIVED October 7, 1988

Author Index

Affiliation Index

Subject Index

Production by Barbara J. Libengood
Indexing by Janet S. Dodd

Elements typeset by Hot Type Ltd., Washington, DC
Printed and bound by Maple Press, York, PA